SONOCHEMISTRY

New Opportunities
for Green Chemistry

SONOCHEMISTRY

New Opportunities for Green Chemistry

Gregory Chatel

Université Savoie Mont Blanc, France

World Scientific

NEW JERSEY · LONDON · SINGAPORE · BEIJING · SHANGHAI · HONG KONG · TAIPEI · CHENNAI · TOKYO

Published by

World Scientific Publishing Europe Ltd.

57 Shelton Street, Covent Garden, London WC2H 9HE

Head office: 5 Toh Tuck Link, Singapore 596224

USA office: 27 Warren Street, Suite 401-402, Hackensack, NJ 07601

Library of Congress Cataloging-in-Publication Data
Names: Chatel, Gregory.
Title: Sonochemistry : new opportunities for green chemistry / Gregory Chatel
    (Université Savoie Mont Blanc, France).
Description: New Jersey : World Scientific, 2016. | Includes bibliographical references.
Identifiers: LCCN 2016023266| ISBN 9781786341273 (hc : alk. paper) |
    ISBN 9781786341501 (pbk : alk. paper)
Subjects: LCSH: Sonochemistry. | Green chemistry. | Ultrasonic waves. |
    Chemistry, Physical and theoretical.
Classification: LCC QD801 .C43 2016 | DDC 541--dc23
LC record available at https://lccn.loc.gov/2016023266

British Library Cataloguing-in-Publication Data
A catalogue record for this book is available from the British Library.

Desk Editors: Anthony Alexander/Mary Simpson

Typeset by Stallion Press
Email: enquiries@stallionpress.com

Foreword

The publication of this new textbook on sonochemistry is a welcome addition to the chemistry literature and a sign that the subject continues to flourish. The book highlights the links between sonochemistry and green chemistry, a relationship that was identified many years ago but has been rejuvenated over the last few years. Let us explore how the connection was first established between sonochemistry and green chemistry.

When I began my own research into the effects of ultrasound on chemical reactions nearly 40 years ago the term "sonochemistry" did not exist in the scientific literature. In those days my colleague Phil Lorimer and I had to fight a long battle with other chemists in the UK and elsewhere to convince them that ultrasound could indeed accelerate chemical reactions. At the same time, as we were beginning our work and isolated from us there were several groups who had begun independent investigations of various ultrasonic effects on chemical reactions in areas such as polymerization (P. Kruus) organometallics (J.-L. Luche, P. Boudjouk, K. Suslick) and more general effects of acoustic cavitation (A. Henglein, P. Riesz, M. A. Margulis). Together with other researchers these studies began to be drawn together in the 1980s and the subject became known as sonochemistry. The new name was certainly a positive move in that the title could be used as an "umbrella" to gather together scientists working on the effects of ultrasound over a wide range of chemistry and processing. In later years, however, the term sonochemistry proved to be a little restrictive because it suggested that it applied only to topics related to chemistry even though the subject was expanding dramatically to include food science, microbiology, environmental protection, materials and therapy. In those early days Arnim Henglein, who had been working with cavitation for many years before the 1980s, warned that the term "sonochemistry" could be misleading because the subject was based on the effects of cavitation and should involve other disciplines such as physics, engineering and medicine. Nevertheless the name has now become firmly estabished and is unlikely to be changed.

The original focus of sonochemistry was synthesis and its general acceptance as a new methodology led to the formation of the European Society of Sonochemistry. At its first meeting in Autrans (1990) some 75% of the presentations related to chemical synthesis and these together with other sonochemical synthesis papers published around that time concentrated on the way in which ultrasound could improve or accelerate reactions. It became clear in subsequent years that sonochemistry had an important role to play in the emerging field of green chemistry (T. J. Mason and S. S. Phull, "Sonochemistry in waste minimisation", In: *Chemistry of Waste Minimisation*, Ed. J. H. Clark, Springer Netherlands 1995, 328–359, T. J. Mason and P. Cintas, "Sonochemistry", In: *Handbook of Green Chemistry and Technology*, Eds. J. H. Clark and D. Macquarrie, Wiley-Blackwell, 2002, 372–396). At that time the basic descriptions of Green Chemistry and of Green Engineering were similar:

Green Chemistry: the design of chemical products and processes that reduce or eliminate the use and generation of hazardous substances and seek to reduce and prevent pollution at source.
Green Engineering: the design, commercialization and use of processes that are feasible and economic, reduce the generation of pollution at the source and minimize the risk to human health and the environment.

When we compare these definitions with some of those that were used to describe sonochemistry then a remarkable similarity becomes evident (T. J. Mason, Sonochemistry and sonoprocessing: the link, the trends and probably the future, *Ultrason. Sonochem.* **2003**, *10*, 175–179).

Sonochemistry involves:

- The use of less hazardous chemicals and environmentally friendly solvents.
- Developing reaction conditions that increase the selectivity for the product.
- Minimizing the energy consumption for chemical transformations.
- The possibility of using alternative or renewable feedstocks, in particular biomaterials.

This book has taken that story forward to the present day and shown how sonochemistry can be used to introduce new approaches to green chemistry.

Perhaps more importantly for me is that this book helps provide some assurance that there is a healthy future for my subject. It is also comforting for someone who helped to develop sonochemistry that it is now in the capable hands of younger researchers such as Gregory Chatel who will actively explore, extend and broaden the many and varied applications of this very wide-ranging science.

Prof. Em. Timothy J. Mason
Faculty of Health and Life Sciences
Coventry University (UK)

Preface

When we talk about ultrasound, we directly think about fetus image scans, or echolocation of bats and dolphins… but you can also find ultrasound in chemistry! More precisely, **power ultrasound** is used in chemistry. In fact, we call that **sonochemistry**, in contrast to ultrasonography, involving **diagnostic ultrasound** used in both veterinary and human medicine.

Amazing to use sound waves to make chemistry! However, some chemists performed the first experimental sonochemical reactions in 1927, even if the term "sonochemistry" was used for the first time in 1980. I cannot forget to note in this preface the remarkable achievements of the French sonochemist Jean-Louis Luche, who defined the "true sonochemistry" and the "false sonochemistry". Jean-Louis Luche, who left us in 2014, said that *"success in sonochemistry relies both on the quality of the equipment and expertise with its use benefit from the full potential offered by ultrasound in the field of chemistry"*. It is totally right and I can add that *"the scientific rigor is essential in the area of sonochemistry to understand the associated mechanisms and benefit from the full potential offered by ultrasound"*. Indeed, the use of ultrasound is often considered as a simple tool of mixing, rather than a **real technology of unconventional activation**.

That is why the goal of this handbook is to introduce sonochemistry, particularly to students, to those wishing to start in the field, or even to researchers wishing to better understand how ultrasonic technology could help them in their particular field of chemistry. The objectives are to give you an **overall introduction of sonochemistry** without entering the details of the acoustic physics, giving **basic theoretical and practical considerations** on the use of ultrasound in labs, to better understand the mechanisms involved and to know which parameters can affect the observed results. In addition, **numerous examples** are reported to highlight how sonochemistry can represent an innovative way to make some chemical processes more eco-friendly and more eco-efficient, in connection with the 12 principles of green chemistry.

After an introductory section to define sonochemistry, sound wave properties, the history of ultrasound and some words on green chemistry, a very tutorial section will deal with the acoustic cavitation (bubble dynamic, physical and chemical effects induced by ultrasound and all the factors affecting cavitation). The latter part will allow the reader to understand, in a simple way, the theory to know how to start operating the ultrasound. Then, it will be essential to explain which parameters are necessary to calculate or estimate to discuss the efficiency of ultrasound assisted processes. Section 4 will be about the laboratory equipment and the opportunity to scale up the technology for developing continuous flow processes. Section 5 will present many applications of sonochemistry for green chemistry through several fields of chemistry. It allows illustrating with precise and recent examples how sonochemistry can improve some reactions or processes according to the 12 principles of Anastas and Warner. The conclusive part will highlight the current limitations of sonochemistry and challenges for research in the area. This critical and personal part will discuss how sonochemistry can be great for green chemistry to conclude the handbook.

I would like to warmly thank Prof. Tim Mason for providing this foreword and the international experts in the area who have agreed to share their opinion with the readers and provide an added value to this handbook. Please appreciate this famous final part as the exciting conclusion of this handbook!

Dr. Gregory Chatel
Université Savoie Mont Blanc
gregory.chatel@univ-smb.fr

About the Author

Dr. Gregory Chatel received his PhD degree in 2012 from the Université de Grenoble (France) under the supervision of Prof. M. Draye and Prof. B. Andrioletti. During his PhD, he particularly developed and fundamentally studied a sonochemical method involving ionic liquids for the epoxidation of various alkenes. In 2013, he joined Prof. R. D. Rogers' group at The University of Alabama (USA) at the Center for Green Manufacturing as a postdoctoral research fellow. His research was focused on the application of ionic liquids in green chemistry, separation and biomass processing. At the end of 2013, Dr. Chatel joined the Institut de Chimie des Milieux et Matériaux de Poitiers (IC2MP) as an Assistant Professor of the Université de Poitiers (France) to develop a biomass valorization program based on non-conventional media/techniques, in particular based on sonochemistry. In 2014, he became the first president of the French national Young Chemists' Network (RJ-SCF) of the French Chemical Society (SCF). In 2016, he joined the Laboratoire de Chimie Moléculaire et Environnement (LCME) of the Université Savoie Mont Blanc (France), the laboratory where Jean-Louis Luche developed the organic sonochemistry that constitutes the bases of one of the aspects of the current green chemistry.

Contents

Chapter 1

Introduction

1. What is the Sonochemistry?

The term *"sonochemistry"* is used to describe the chemical and physical processes occurring in solution through the energy brought by power ultrasound.[1–3] The effects of ultrasound are the consequence of the **cavitation phenomenon**, which is the formation, the growth and the collapse of gaseous microbubbles in liquid phase (Figure 1).[3–5] Ultrasound is propagated through a series of compression and rarefaction waves in the liquid medium. When the acoustic power is sufficiently high, the rarefaction cycle exceeds the attractive forces of the molecules of the liquid and cavitation bubbles of a few micrometers in diameter are formed. Small amounts of vapor or gas from the medium enter in the bubble during its expansion phase and is not fully expelled during compression phase. The bubbles grow over the period of a few cycles to an equilibrium size for the particular frequency applied. The **intense local effects (mechanical, thermal and chemical)** due to the sudden collapse of those bubbles are at the origin of all the applications of sonochemistry.[6–8]

In water, at an ultrasonic frequency (f) of 20 kHz, each cavitation bubble collapse represents a **localized hot-spot**, generating temperatures of about 5,000 K and pressures superior to 1,000 bars (Figure 1). Many factors can affect the cavitation and the results of a sonochemical reaction: the acoustic power, the frequency, the hydrostatic pressure, the nature and the temperature of the solvent, the used gas and even the geometry of the reactor.[9-13] The potential of sonochemistry is often directly connected to the choice of the sonochemical parameters or experimental conditions.

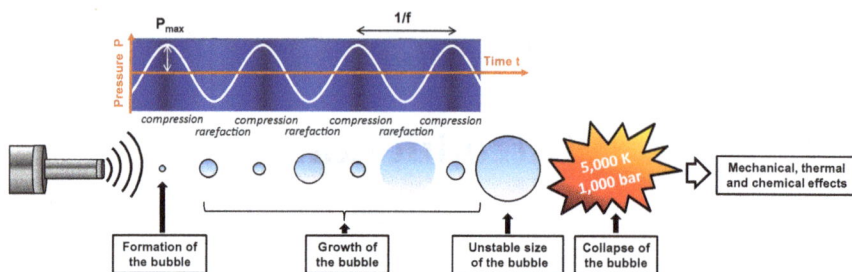

Figure 1: Schematic representation of the acoustic cavitation phenomenon.

For example, the **frequency** is an essential parameter (Figure 2). Indeed, even if the whole mechanism is not elucidated yet, it is usually accepted that, in water, **low frequencies (20–80 kHz)** lead preferentially to physical effects (shockwaves, microjets, microconvection, etc.). On the contrary, **high ultrasonic frequencies (150–2,000 kHz)** favor the production of hydroxyl radicals (HO·) through the local hotspots produced by cavitation, mainly leading to chemical effects. In a broad outline, it is possible to identify two great families of power ultrasound applications in chemistry based either on sonophysical effects or sonochemical effects. Conditions obtained in a medium submitted to ultrasound are accountable for a large number of physicochemical effects as increase in kinetics of chemicals reactions, changes in reaction mechanisms, emulsification effects, erosion, crystallization, precipitation, etc.[14,15]

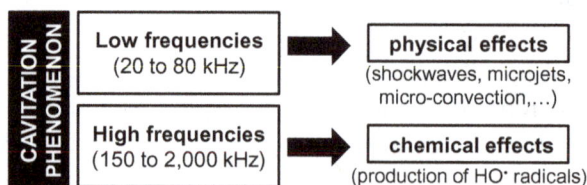

Figure 2: Predominant effects in water as a function of the frequency range.

The design of organic reactions, material preparations or other chemical processes under ultrasound requires **a rigorous methodology** and **the complete report of all sonochemical parameters and experimental details**. This latter point is essential for two reasons: (1) based on the available literature, the comparison between the obtained results and the wide spread of mechanistic explanations is not always trivial; (2) in the absence of these precautions, it could be really difficult (and even impossible in many cases) to reproduce an experiment from the literature.[16]

2. A Chemistry Based on Ultrasonic Waves

By definition, an ultrasonic wave is a sound wave belonging to the range between **20 kHz and 200 MHz** that can be subdivided into two distinct regions: power ultrasound and diagnostic ultrasound (Figure 3). At lower frequencies, greater acoustic energy can be generated to induce cavitation in liquid medium (sonochemistry). Ultrasonic frequencies above 2 MHz do not produce cavitation. This range is particularly used in medical imaging (diagnostic ultrasound).

The sound is a wave generated by a mechanical vibration that travels due to **the elasticity of the surrounding environment** through longitudinal waves (straight line).

⚠ The sound waves do not propagate in vacuum.
⚠ The sound propagation is an elastic deformation (reversible): there is no transport of material.

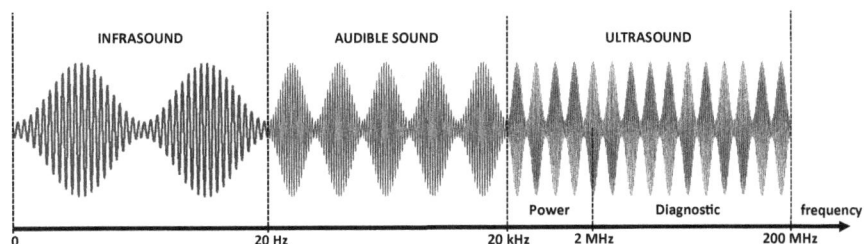

| INFRASOUND | AUDIBLE SOUND | ULTRASOUND |

Power | Diagnostic | frequency

0 | 20 Hz | 20 kHz | 2 MHz | 200 MHz

Figure 3: Frequency ranges of sound.

3

To be described, a sound wave is often considered as a **sinusoidal plane wave** which is particularly characterized by the following properties:

- The **frequency** (f) in Hertz (Hz) or its inverse, the **period** (T) in seconds (s):

$$f = \frac{1}{T}. \tag{1}$$

- The **wavelength** (λ) usually expressed in nanometers (nm) or its inverse, the **wave number** ($\bar{v}$) in cm^{-1}:

$$\bar{v} = \frac{1}{\lambda}. \tag{2}$$

- The alternative pressure wave is also characterized by its **amplitude** (P). The temporal evolution of the amplitude ($P(t)$) follows the simplified Eq. (3):

$$P(t) = P_{max} \times \sin(2\pi t + \Phi), \tag{3}$$

where P_{max} is the maximal amplitude, t is the time and ϕ is the phase.

Figure 4 schematically represents the wave with the evolution of the amplitude as a function of the time. We do not give more details on the determination/calculation of the parameters such as amplitude or phase since they should not be very useful for the chemist who wants to develop sonochemical applications.

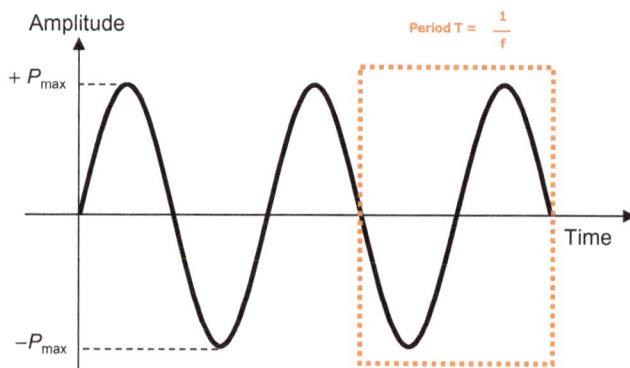

Figure 4: Representation of the sinusoidal plane wave of sound.

- The **sound pressure** or **acoustic pressure**, expressed in Pascal (Pa), is the local pressure deviation from the ambient pressure caused by a sound wave. It could be measured using a microphone in air or a hydrophone in water.
- The **speed of sound**, denoted c and expressed in meter per second $(m \cdot s^{-1})$, is the distance traveled per unit time by a sound wave propagating through an elastic medium.

As this handbook is not focused on the physics of waves, but rather on the use and applications of ultrasound for chemistry, we can recommend to you the recent book written by Prof. Tim Freegarde from the University of Southampton.[17]

3. The Brief History of Sonochemistry

The history of sonochemistry is relatively recent and to better understand, we have to date back to the discovery of ultrasound.

Indeed, this story starts with the Italian biologist Lazzaro Spallanzani (1729–1799) who discovered in 1794 by binding bat eyes that their movements were directed thanks to ultrasound: in fact, their eyes were their ears! However, ultrasound was really discovered in 1883 by the English physiologist Sir Francis Galton (1822–1911) who invented the silent whistle or "Galton's whistle", emitting sound in a range that only dogs heard, but not humans: this range of sound was ultrasonic range![18]

The discovery of the piezoelectricity by French physicists Jacques Curie (1855–1941) and Pierre Curie (1859–1906) allowed the generation of ultrasound in water from piezoelectric materials and some powerful electronic devices.

After the Titanic tragedy of 1912, Paul Langevin (1872–1946) suggested the use of ultrasound for the detection of icebergs. In 1917, he developed a sonar system, called hydrophone, using the ultrasonic vibrations to detect submarines by echolocation. This represents the first industrial application of ultrasound during the World War.

As a fundamental point of view, the cavitation phenomenon was first observed by Sir John Isaac Thornycroft (1843–1928) and Sydney Walker Barnaby (1855–1925) with the propeller of their submarine becoming

Piezoelectric effect:

Piezoelectricity is the property of certain materials to electrically polarize under the action of mechanical force and reversely, to deform when an electric field is applied to them. Both effects are inseparable.

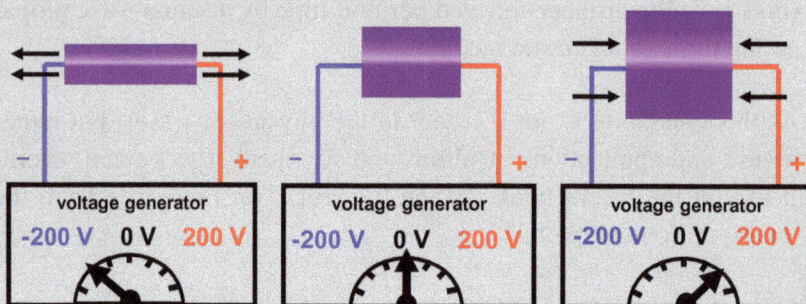

As shown above, when a potential difference is applied between two faces of the discs of a piezoelectric material (for example crystals of quartz), it expands or contracts with an opposite effect if the potential difference between the two sides is reversed. If variable potential difference is applied in time, there will be a succession of contractions–expansions phases following the changes of potential differences.

pitted and eroded over a relatively short operation period.[19] They reported *"unusual vibrations of their propeller due to large bubbles generated by the movement of the blades"* and *"the implosion of these bubbles under the water pressure."* Lord Rayleigh (1842–1919) explained the erosion of the propellers of boats by the formation and growth of vapor bubbles in the presence of a depression caused by Bernoulli effect, followed by a violent collapse. He determined the first mathematical model describing a cavitation event in an incompressible fluid.[20]

The story of sonochemistry really started in 1927 when Alfred Lee Loomis (1887–1975), Robert William Wood (1868–1955) and Theodore William Richards (1868–1928) reported for the first time chemical and biological effects of ultrasound, showing that cavitation could be a useful tool in chemistry.[21,22] One of the first applications was the use of ultrasound inducing cavitation to degrade a biological polymer.[23] The use of ultrasound was industrially developed in the 1950s for cleaning applications in heavy industry, medical instruments, clothing and textiles and food industry.[24]

However, it took until the 1980s and the onset of reliable and marketed ultrasonic generators for the researchers to demonstrate that the ultrasonic waves offer undeniable opportunities in chemistry. In 1980, Ernest Arthur Neppiras introduced the term "sonochemistry" in a review about acoustic cavitation.[25] From that moment, progress in the field of sonochemistry increased exponentially. Timothy J. Mason particularly promoted the use of ultrasound in various areas of chemistry, through fundamental aspects as well as innovative applications,[26–28] and by co-founding in 1991 the European Society of Sonochemistry and a new Elsevier® journal dedicated to sonochemistry works (*Ultrasonics Sonochemistry*). In 1998, Jean-Louis Luche, also considered as a pioneer of modern sonochemistry, described the true and false effects of sonochemistry (see Section 1 in Chapter 5, page 59).[29] The major part of this handbook is dedicated to understand why sonochemistry can be considered as a beneficial tool for green chemistry.

4. What about Green Chemistry?

In 1990, the Environmental Protection Agency (EPA) published the *Pollution Prevention Act*, which clearly ruled that the prevention of pollution by source reduction is the most desirable level of environmental protection.[30] Paul T. Anastas, chemist and member of the EPA, actively participated in these discussions and published in 1996 with Tracy C. Williamson a book entitled *Green Chemistry: Designing Chemistry for the Environment*, establishing the basis of Green Chemistry.[31]

In 1998, Paul T. Anastas and John C. Warner published the 12 principles of green chemistry, the first concrete reflection tool for a more sustainable and eco-friendly chemistry.[32,33] This text is considered as the founding act of green chemistry: *"Green chemistry is the utilization of a set of principles that reduces or eliminates the use or generation of hazardous substances in the design, manufacture and applications of chemical products."*[32]

The 12 principles of Green Chemistry[32]:

1. Prevention;
2. Atom economy;
3. Less hazardous chemical syntheses;
4. Designing safer chemicals;
5. Safer solvents and auxiliaries;
6. Design for energy efficiency;
7. Use of renewable feedstocks;
8. Reduce derivatives;
9. Catalysis;
10. Design for degradation;
11. Real-time analysis for pollution prevention;
12. Inherently safer chemistry for accident prevention.

Green chemistry is currently based on five basic concepts:

⇨ **Prevention:** it is better to prevent waste than to treat or clean up waste afterwards.

⇨ **Better use of the raw material:** atom economy concept takes into account the optimal transformation of the raw material in the final product, limiting the production of by-products. Renewable feed-stocks are clearly favored.

⇨ **Better wastes management:** produce minimal amount of waste and/or in an appropriate form which limits the potential spread and facilitate their recycling.

⇨ **Energy savings:** energy has to be better managed in terms of sources, savings and efficiencies.

⇨ **Use of solvents compatible with the environment:** some toxic solvents for humans and the environment have to be abandoned. The use of supercritical fluids, ionic liquids and/or water as solvents or even the development of processes using no solvent (processes related to the chemistry of microwaves for example) are brought forward as alternatives to the use of volatile and toxic organic solvents.

In 2003, Paul T. Anastas and Julie B. Zimmerman developed the 12 principles of Green Engineering to outline what would make a greener chemical process or product for scale-up at an industrial level.[34] All process steps (such as steps of purification and separation) and the afterlife of products (life cycle analysis, for example) are considered through these 12 supplementary principles.[35] For more details on green chemistry and engineering, several recent interesting books that focus on the introduction of these themes can be referred.[36–40]

The 12 principles of Green Engineering[34]:

1. Designers need to strive to ensure that all material and energy inputs and outputs are as inherently non-hazardous.
2. It is better to prevent waste than to treat or clean up waste after it is formed.
3. Separation and purification operations should be designed to minimize energy consumption and materials use.
4. Products, processes and systems should be designed to maximize mass, energy, space and time efficiency.
5. Products, processes and systems should be "output pulled" rather than "input pushed" through the use of energy and materials.
6. Embedded entropy and complexity must be viewed as an investment when making design choices on recycle, reuse or beneficial disposition.
7. Targeted durability, not immortality, should be a design goal.
8. Design for unnecessary capacity or capability (e.g., "one size fits all") solutions should be considered a design flaw.
9. Material diversity in multicomponent products should be minimized to promote disassembly and value retention.
10. Design of products, processes and systems must include integration and interconnectivity with available energy and materials flows.
11. Products, processes, and systems should be designed for performance in a commercial "afterlife".
12. Material and energy inputs should be renewable rather than depleting.

As we will show in this handbook, the relationships between green chemistry, green engineering and sonochemistry are very narrow, especially in terms of processes, energy efficiencies and applications.

References

1. T. J. Mason, *Sonochemistry*, Oxford University Press, Oxford, UK, 1999, 96.
2. T. J. Mason, *Advances in Sonochemistry*, Vol. 5, Elsevier, Jai Press Inc. Stamford, Connecticut, 1999, 310.
3. J.-P. Bazureau, M. Draye, *Ultrasound and Microwaves: Recent Advances in Organic Chemistry*, Transworld Research Network, Kerala, 2011.
4. K. S. Suslick, D. A. Hammerton, D. E. Cline, *J. Am. Chem. Soc.* **1986**, *108*, 5641–5645.
5. T. J. Mason, D. Peters, *Practical Sonochemistry: Power Ultrasound Uses and Applications*, 2nd Ed., Woodhead Publishing, Cambridge, UK, 2002, 166.
6. T. J. Mason, *Practical Sonochemistry: User's Guide to Applications in Chemistry and Chemical Engineering*, Ellis Horwood Ltd, New York, 1992.
7. J. L. Luche, *Synthetic Organic Sonochemistry*, Plenum Press, New York, 1998.
8. P. R. Gogate, A. B. Pandit, Sonocrystallization and its application in food and bioprocessing, In: *Ultrasound Technologies for Food and Bioprocessing*, (Eds: H. Feng, G. V. Barbosa-Cánovas, J. Weiss) Food Engineering Series, New-York, USA 2011, 467–493.
9. A. Henglein, M. Gutierrez, *J. Phys. Chem.* **1993**, *97*, 158–162.
10. T. J. Mason, *Sonochemistry*, Chemistry Primers, Oxford, 2000.
11. S. I. Nikitenko, C. Le Naour, P. Moisy, *Ultrason. Sonochem.* **2007**, *14*, 330–336.
12. S. de La Rochebrochard d'Auzay, J.-F. Blais, E. Naffrechoux, *Ultrason. Sonochem.* **2010**, *17*, 547–554.
13. P. R. Gogate, P. A. Tatake, P. M. Kanthale, A. B. Pandit, *AIChE J.* **2002**, *48*, 1542–1560.
14. T. J. Mason, *Ultrason. Sonochem.* **2003**, *10*, 175–179.
15. G. Cravotto, P. Cintas, *Angew. Chem. Int. Ed.* **2007**, *46*, 5476–5478.
16. T. J. Mason, E. Cordemans de Meulenaer, Practical considerations for process optimisation, In: *Synthetic Organic Sonochemistry*, (Ed.: J.-L. Luche) Plenum Press, New York, 1998, 301–328.
17. T. Freegarde, *Introduction to the Physics of Waves*, Cambridge University Press, Cambridge, UK, 2013.
18. F. Galton, *Inquiries into Human Faculty and its Development*, Macmillan, London, UK, 1883.
19. J. Thorneycroft, S. W. Barnaby, *Inst. Civil Eng.* **1895**, *122*, 51–55.
20. L. Rayleigh, *P. Mag.* **1917**, *34*, 94–98.
21. R. Woods, A. Loomis, *Philos. Mag.* **1927**, *4*, 414–436.

22. T. Richards, A. Loomis, *J. Amer. Chem. Soc.* **1927**, *49*, 3086–3100.
23. S. Brohult, *Nature* **1937**, *140*, 805–810.
24. T. J. Mason, *Ultrason. Sonochem.* **2016**, *29*, 519–523.
25. E. Neppiras, *Phys. Rep.* **1980**, *61*, 159–284.
26. T. J. Mason, J. P. Lorimer, *Sonochemistry, Theory, Applications and Uses of Ultrasound in Chemistry*, Ellis Horwood Publishers, Chichester, 1988.
27. T. J. Mason, *Chem. Soc. Rev.* **1997**, *26*, 443–451.
28. T. J. Mason, J. P. Lorimer, *Applied Sonochemistry*, Wiley-VCH Verlag GmbH, Weinheim, 2002.
29. J.-L. Luche, *Synthetic Organic Sonochemistry*, Plenum Press, New York, 1998.
30. EPA, *Pollution Act of 1990*, United States Code Title 42, 1990.
31. P. T. Anastas, T. C. Williamson, Green chemistry: An overview, In: *Green Chemistry*, American Chemical Society, Washington, 1996, 1–17.
32. P. T. Anastas, J. C. Warner, *Green Chemistry: Theory and Practice*, Oxford University Press, Oxford, 1998.
33. P. T. Anastas, N. Eghbali, *Chem. Soc. Rev.* **2010**, *39*, 301–312.
34. P. T. Anastas, J. B. Zimmerman, *Env. Sci. Technol.* **2003**, *37*, 94A–101A.
35. P. T. Anastas, J. B. Zimmerman, *Sustainability Science and Engineering Defining Principle*, Elsevier, Amsterdam, 2006.
36. P. T. Anastas, R. Boethling, C.-J. Li, A. Voutchkova, A. Perosa, M. Selva, *Handbook of Green Chemistry*, Vol. 3, Wiley-VCH, Weinheim, 2012.
37. R. A. Sheldon, I. Arends, U. Hanefeld, *Green Chemistry and Catalysis*, Wiley-VCH, Weinheim, 2007.
38. R. Luque, J.-C. Colmenares, *An Introduction to Green Chemistry Methods*, Future Science Ltd, 2013.
39. A. Matlack, *Introduction to Green Chemistry*, 2nd Ed., CRC Press, USA, 2010.
40. J. González, D. J. C. Constable, *Green Chemistry and Engineering: A Practical Design Approach*, Wiley, Hoboken, New Jersey, 2011.

Chapter 2

Acoustic Cavitation

1. Bubble Dynamic

As described in Figure 1 (page 2), the cavitation phenomenon starts with the formation of gaseous microbubbles in liquid phase and its radial oscillation. In fact, the expansion cycles exert negative pressure on the liquid, pulling the molecules away from one another. If the ultrasonic wave is sufficiently intense, the expansion cycle creates cavities in the liquid. Then, the negative pressure exceeds the local tensile strength of the liquid, which varies according to the type and purity of liquid.[1,2]

Note: Cavitation is a nucleated process. It occurs at pre-existing weak points in the liquid, such as gas-filled crevices in suspended particulate matter or transient microbubbles from prior cavitation events. Most liquids are sufficiently contaminated by small particles such that cavitation can be readily initiated at moderate negative pressures.

Once formed, small irradiated gas bubbles absorb energy from the sound waves and grow (Figure 1). Cavity growth depends on the intensity of the sound.

At high intensities, a small cavity may grow rapidly through inertial effects. If cavity expansion is sufficiently rapid during the expansion half of a single cycle, it will not have time to recompress during the compression half of the acoustic cycle.

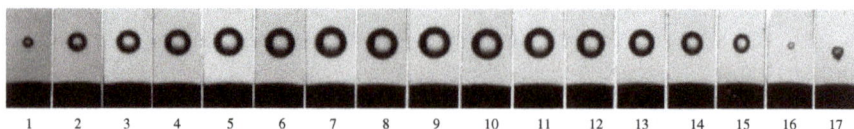

Figure 1: High-speed photographs of growth and collapse of cavitation bubbles (Frame interval 10 μs, exposure 2 μs). Reproduced with the permission of Ref. 3.

At lower acoustic intensities, the cavity oscillates in size over many expansion and compression cycles. During such oscillations, the amount of gas or vapor that diffuses in or out of the cavity depends on the surface area, which is slightly larger during expansion than during compression (Figure 2). Cavity growth during each expansion is slightly larger than shrinkage during the compression. Thus, the cavity will grow over many acoustic cycles. The growing cavity can eventually reach a critical size where it can efficiently absorb energy from the ultrasonic irradiation. This critical size depends on the liquid and the frequency of sound. At this point, the cavity can grow rapidly during a single cycle of sound.

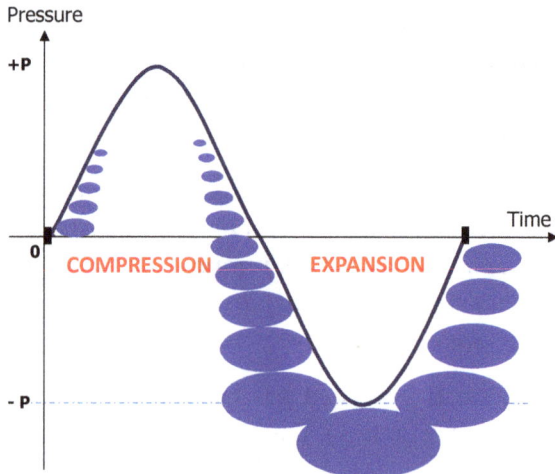

Figure 2: Schematic representation of bubble dynamic through one oscillation showing a compression phase and an expansion phase.

Once the cavity has overgrown, it can no longer absorb energy as efficiently.[4] Without the energy input, the cavity can no longer sustain

Figure 3: Picture of liquid jet formation during cavitation bubble collapse. Reprinted with permission from Ref. 6. Copyright (2016) Elsevier.

itself. The surrounding liquid rushes in and the cavity implodes (Figure 3). This sudden collapse of the cavity creates an unusual environment for chemical reactions releasing significant amounts of energy in the form of intense local heat (up to 5,000 K), very high pressure (close to 1,000 bar), divergent shock waves at immediate surrounding of the bubble, acoustic microcurrents and violent liquid microjets (up to 100 m·s^{-1}). These local effects (mechanical, thermal and chemical) are at the origin of all the applications of sonochemistry.[5]

To summarize, **bubble dynamic** is, as a first approximation, the result of the competition of two inertial forces: cohesion forces resulting from superficial tension and the oscillating pressure. Some specific publications explain more fundamentally the theory of this phenomenon and report the associated (and complex!) equations.[7–9]

2. Cavitation, Effects and Models

Cavitation is defined as a disturbance of the continuous liquid medium under the effect of excessive stress accompanied by the formation, the growth and finally the violent implosion of bubbles created by this disturbance (Figure 1, page 2).

Interactions between gas bubbles and ultrasonic waves depend on the amplitude of the pressure variation. At low amplitude, gas bubbles vary linearly with pressure variations. For each variation of pressure, a resonance frequency corresponds to the maximal of the vibration of the bubble. This **stable cavitation** does not usually lead to any sonochemical phenomena.[10]

However, at higher amplitude of pressure variation, the response of the gas bubble can become nonlinear. Indeed, the growth of bubbles during the expansion phase (Figure 2) leads to an increase of the gaseous volume that is superior to the decrease taking place during the compression step and bubbles shrink with a very high speed on a small volume in a sudden collapse, causing the release of extreme temperatures and pressures. These bubbles undergo asymmetric distortions projecting liquid jets into the bubbles (Figure 4). This **transient cavitation** is at the basis of the different physico-chemical effects used in sonochemistry. In addition, the generated high temperatures can cause the dissociation of the solvent vapors or other gaseous molecules into radicals (such as $H^{\bullet}$ and $HO^{\bullet}$ *via* the sonolysis of water) as well as the emission of light (see Section 9 in Chapter 3, page 34).[11]

As a fundamental point of view, we currently find four theories to explain sonochemical effects: (1) hot-spot theory[12]; (2) electrical theory[13,14]; (3) supercritical theory[15] and (4) plasma discharge theory.[16–18] Even if it does not explain all the sonochemical phenomenon (see Section 1 in Chapter 6, page 147, for more details), the **hot-spot theory** is the most widely accepted in explaining sonochemical processes.[5,19–21] According to this theory, each formed microbubble acts as a microreactor which produces different reactive species and heat during its collapse. The temperature profile shows

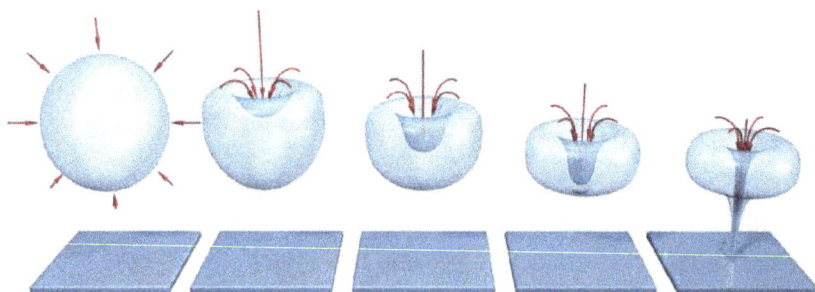

Figure 4: Microjet creation through collapsing cavitation bubbles (ask permission from website: http://eswt.net/cavitation).

Source: Courtesy of FOCUS-IT, LLC

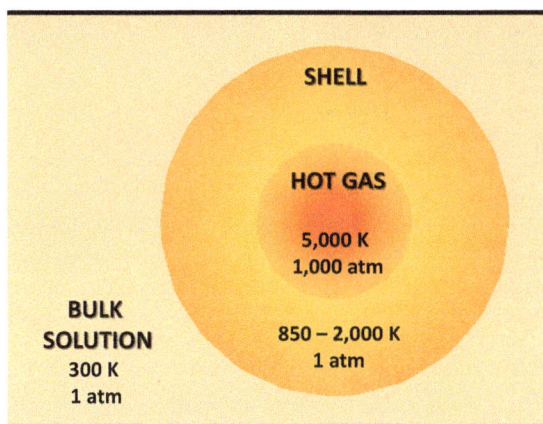

Figure 5: The three reactive zones associated with a cavitational bubble according to the hot-spot theory.

that there are three zones associated with a cavitational bubble (Figure 5): **(a) the core of the bubble in the gaseous phase; (b) the interface shell between gaseous and liquid phases and (c) the liquid bulk solution.**

At the core of the cavitation bubble, where extreme conditions (temperatures of about 5,000 K and pressures of 1,000 atm) are found, solvent can be pyrolyzed to form radical species (sonolysis of water, for example). In the sonochemical reactions, the substrates with a great volatility or with a poor degree of solvation can directly react at the core of the cavitation bubble with the formed radicals or undergo pyrolysis.[4,22]

The reactions **at the interfacial region (the shell)** between gaseous phase and bulk liquid correspond to an indirect mechanism wherein the sonolysis of the solvent or volatile solutes in the bubble constitutes a first step. The temperature of this layer to the boundary of the bubble was estimated between 850 K and 2,000 K depending on the studies.[23] In aqueous solutions, the hydrophobic compounds are more concentrated in this region than in the bulk solution.

The third reactive zone is the liquid bulk solution where the substrates which react are poorly volatile or non-volatile and strongly solvated. In aqueous solution, the oxidation products present similarities with the products obtained by radiolysis or advanced oxidation processes (H_2O_2/UV, O_3/UV, O_3/H_2O_2, etc.).[24,25] In this case, the mechanism is based on the radical attacks by the $HO^\bullet$. In the bulk region, the temperature remains at a level similar to room temperature because cavitation is an adiabatic process.

Incredible effects of sonochemistry through the hot-spot![26]

For a rough comparison, the collapse of the cavitation bubbles leads to incredible effects by reaching the temperature of the surface of the sun, the pressure at the bottom of the ocean, the lifetime of a lightning strike and a million times faster cooling than a red hot iron rod plunged into water!

We can distinguish the effects of ultrasound in homogeneous and heterogeneous media. For homogeneous media, sonochemical reactions are related to new chemical species produced during cavitation, whereas for the latter, enhancement of the reactions could also be related to mechanical effects induced in liquid system by sonication. Several examples in homogeneous and heterogeneous conditions will be detailed in the chapter of this handbook related to the applications in the area of green chemistry (see Chapter 5, page 59).

3. Factors Affecting Cavitation

The ambient conditions of a reaction system can strongly influence the intensity of the acoustic cavitation that can directly affect the rate or the yield of the chemical reaction. Indeed, the occurrence of acoustic cavitation in the liquid medium can be dependent on sonochemical parameters (frequency, acoustic power, geometry of the used reactor) and experimental conditions (pressure, type of solvent, temperature, presence or not of solid impurities or dissolved gasses, etc.). Even if all the mechanisms are not clearly demonstrated, this section provides the main trends in the evolution of sonochemical effects as a function of the chosen parameters.

3.1. Frequency

The ultrasonic frequency (noted f, expressed in Hz) determines the short lifetime of the bubbles during the transient cavitation. The maximum radius of the bubbles decreases proportionally with the frequency of the ultrasonic wave, according to a ratio $3/f$. This maximum radius at which bubbles lose their stability corresponds to the resonance radius, determined for a specific frequency, called Minnaert frequency.[27]

It is usually accepted that, in water, **low frequencies (20–80 kHz)** allow obtaining few bubbles with larger dimensions, preferentially leading to **physical effects** (shockwaves, microjets, microconvection, etc.). On the contrary, **high ultrasonic frequencies** (150–2,000 kHz) allow the production of numerous bubbles with smaller diameter, favoring the production of hydroxyl radicals (HO•) through the local hotspots, mainly leading to **chemical effects**. In addition, the expansion time (cycle during the pressure rarefaction) and collapse (during the pressure increasing cycle) is lower at high frequencies than at low frequencies.

Interestingly, Tatake and Pandit investigated the effect of combination of two frequencies by using a combination of two reactors.[28] The combination gave better control over the cavitational activity with enhanced reaction rates due to higher resonance effect on bubble growth as compared to the single frequency operation. Pandit *et al.* proved the higher overall cavitational activity of a triple-frequency sonochemical reactor (30 kHz, 30 kHz, 125 kHz) as compared to the single- and dual-frequency sonochemical reactors at equivalent power dissipation levels.[29] Ashokkumar *et al.* reported that the combined use of low and high frequencies (20 kHz and 355 kHz) decrease the sonochemical efficiency in water. However, a synergetic enhancement of sonochemical rates of hydrogen peroxide *via* the dual-frequency mode was found in the presence of propanol.[30] Wayment and Casadonte described the design of a single-transducer variable-frequency sonication system in the range 20–500 kHz, operating at constant acoustic power. They reported the maximal oxidation of iodide potassium in water at 300 kHz.[31] In the same way, Beckett and Hua determined the best frequency as 358 kHz for the study of the decomposition of 1,4-dioxane.[32] Merouani *et al.* showed the decrease of the degradation rate of acidic diazo dye with the increasing frequency in the range of 585–1,140 kHz.[33]

In reality, we could provide as many trends as applications developed in the literature, particularly since the used equipment is not identical and the applications are different. In addition, **no clear correlation on the frequencies was established for the moment, except that the physical and chemical effects are favored at low and high frequencies, respectively.**

3.2. Acoustic power

The acoustic power is the energy irradiated in the medium in a determined sonication time. It is directly linked to the electric power imposed upon the generator by the operator. A minimum acoustic power is required to observe the phenomenon of cavitation, known as the "Blake threshold", determined from the hydrostatic pressure of the medium as a function of the radius of the cavitation nucleus.[34] The acoustic power also depends on the acoustic impedance of the medium, which increases when the medium starts the process of cavitation. Thus, the temperature and the viscosity of the medium have an influence on the acoustic power.

The evolution of the cavitation as a function of the increase of power is difficult to predict and could be very dependent on the geometry of the reactor or/and the height of irradiated solvent.[35] For example, Son *et al.* found that low power was more effective for the small volume while the large volume required high power level.[36] In the 291 kHz sonoreactor used in this study, a moderate power of $400 \ W \cdot L^{-1}$ was suggested for the most efficient sonochemical oxidation of iodide ion. Kojima *et al.* reported that the acoustic streaming of a liquid increases by increasing the power. However, the sonochemical efficiency (see Chapter 4, page 41) decreases up to a certain electric power.[37]

3.3. Pressure

The more the increase in hydrostatic pressure, the less favorable the cavitation, but the more important the energy liberated by the implosion of the bubble and the corresponding sonochemical effects.[38] Henglein and Gutiérrez observed that the nature of the gas under which the irradiation occurred (1 MHz) had no influence on the sonochemical effects.[39] However, they determined a critical intensity as a function of the pressure. Indeed, the investigation of the pressure (from 0.7 bars to 3 bars) demonstrated that at low intensities, the chemical yield (iodide oxidation) and the luminescence intensity decreased with increasing pressure. Inversely, at higher intensities, the yields increased moderately with increasing pressure.

Some recent examples reported the use of ultrasound under pressure. Raso *et al.* improved the extraction of carotenoids from tomato waste (50 kPa, 6 min, 20 kHz).[41] Delmas *et al.* also investigated sonochemistry

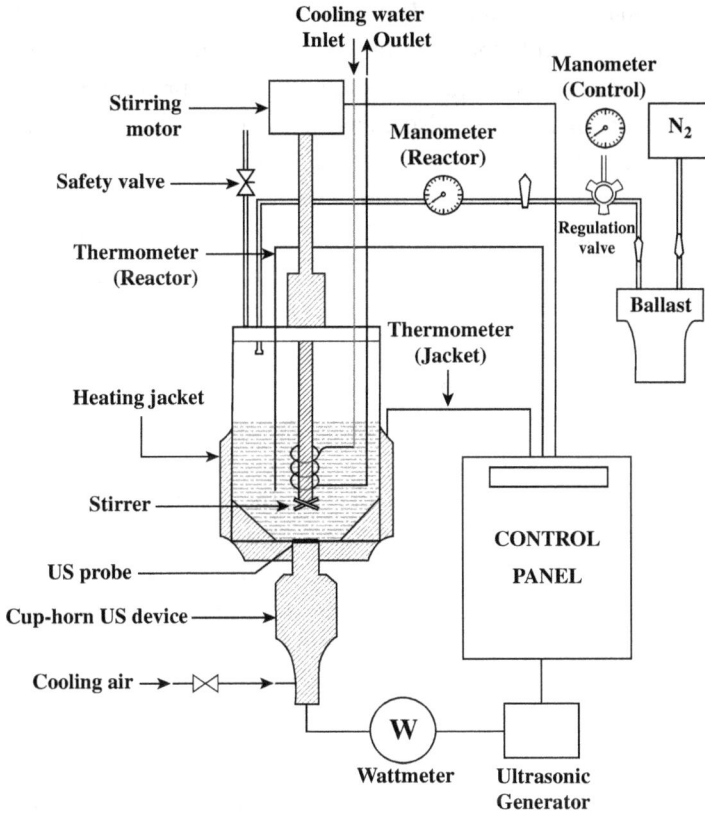

Figure 6: Ultrasonic autoclave implemented by Delmas *et al.* Reprinted with permission from Ref. 40. Copyright (2016) Elsevier.

under pressure for the pretreatment of sludge (Figure 6).[42] The optimum pressure (2 bars) under 20 kHz allowed significant energy savings by reducing the sonication time.

3.4. Temperature

The solvent temperature may play a dual role when it is subjected to ultrasound. On the one hand, increasing the temperature decreases all interactions (Van der Waals forces, hydrogen bonds, dipole attractions, etc.) and improves the diffusion phenomena. On the other hand, cavitation is more easily achieved at lower temperatures when the ultrasonic power of the generator remains constant.[43] Entezari and Kruus explained that the decrease of iodide oxidation as a function of the temperature is due to the vaporous cavitation occurring.[44] Indeed, when the temperature of the liquid increases, its vapor pressure also increases, but much more dramatically than the temperature. The vapor enters the bubble during its collapse and that is why the temperature of the "hot-spot" decreases. As a result, the rate of iodide oxidation decreased as bulk temperature increased at 20 kHz.

In a reasonable range of temperatures (until 80°C), the increase of the temperature of the medium can be interesting to improve the yield of a reaction or the synthesis of materials. For example, Pourabbas *et al.* reported that the ultrasonic preparation of ZnO nanoparticles (20 kHz) was more effective in particle and aggregate average size reduction and narrowness of size distribution at 70°C compared to 25°C.[45]

The cavitation phenomenon causes the continuous increase of temperature of the medium irradiated by ultrasound as a function of the sonication time. It could be important to control this temperature of the medium (with a cooling system) during the irradiation to clearly know the experimental conditions and understand the associated chemistry.

3.5. Solvent

The majority of the ultrasonic applications are performed in water as solvent or liquid of irradiation. However, other solvents can be used under ultrasound, for example organic reactions. The main limitation is that the common organic solvents generally present a high vapor pressure which greatly reduces the intensity of cavitation, limiting the sonochemical effects.[46,47] In addition, the more the cohesive forces acting in the liquid, such as viscosity and surface tension, the more difficult the cavitation phenomenon is to achieve. The viscosity is also closely related to the attenuation of the ultrasonic wave.[48]

3.6. Dissolved gas

Bubbles of dissolved gas in a fluid are able to promote the cavitation through the improvement of the germination phase. For this reason, a gas is often introduced in solution up to saturation by a continuous bubbling to significantly increase the effects of cavitation.[49] The higher the adiabatic and the lower the heat conduction, the higher the hot-spot temperature and the more pronounced are sonochemical effects (radical formation and sonoluminescence). Adding gas to a liquid lowers its surface tension, thereby increasing nucleation but decreasing shape stability of the bubbles.

Conventionally, monoatomic gases such as helium, argon and neon are the most used.[50] Adding molecules that easily form radicals, such as oxygen, can increase radical production, although the hot-spot temperature may be lowered by this addition since the maximum in radical production does not necessarily coincide with the maximum hot-spot temperature.

3.7. Shape of the sonoreactor

Taking into account the low attenuation of ultrasonic waves traveling in a liquid, it is expected that the size and shape of the reactor play a significant role in the topology of the ultrasonic field and as a consequence in sonochemical activity.[51,52]

Interestingly, Naffrechoux *et al.* showed that the effect of liquid height on sonochemical efficiency was more depending on reactor configuration than frequency. The frequency effect that was studied through the low (20 kHz) and high (300–500 kHz) range was shown to be coupled to liquid height effect.[35]

In the case of higher frequencies, the combination of the acoustic streaming with a mechanical stirring could be beneficial. For example, Koda *et al.* reported that the sonochemical efficiency was about twice higher under 490 kHz by introduction of the stirrer.[37]

References

1. F. Grieser, P.-K. Choi, N. Enomoto, H. Harada, K. Okitsu, K. Yasui, *Sonochemistry and the Acoustic Bubble*, Elsevier, Oxford, 2015.
2. T. J. Mason, *Sonochemistry*, Oxford University Press, Oxford, 1999.
3. Y. Tomita, P. B. Robinson, R. P. Tong, J. R. Blake, *J. Fluid Mech.* **2002**, *466*, 259–283.

4. J.-L. Luche, *Synthetic Organic Sonochemistry*, Plenum Press, New York, 1998, 431.
5. K. S. Suslick, D. A. Hammerton, D. E. Cline, *J. Am. Chem. Soc.* **1986**, *108*, 5641–5645.
6. A. J. Coleman, J. E. Saunders, L. A. Crum, M. Dyson, *Ultrason. Med. Biol.* **1987**, *13*, 69–76.
7. C. E. Brennen, *Cavitation and Bubble Dynamics*, Oxford University Press, New York, 1995.
8. M. S. Plesset, A. Prosperetti, *Ann. Rev. Fluid Mech.* **1977**, *9*, 145–185.
9. M. A. Margulis, *Sonochemistry and Cavitation*, Gordon and Breach Publishers, Amsterdam, 1995.
10. T. Lepoint, F. Mullie, *Ultrason. Sonochem.* **1994**, *1*, S13–S22.
11. D. F. Gaitan, L. A. Crum, C. C. Church, R. A. Roy, *J. Acoust. Soc. Am.* **1992**, *91*, 3166–3183.
12. K. S. Suslick, S. J. Doktycz, E. B. Flint, *Ultrason.* **1990**, *28*, 280–290.
13. M. A. Margulis, *Ultrason. Sonochem.* **1994**, *1*, S87–S90.
14. Y. I. Frenkel, *Russ. J. Phys. Chem.* **1940**, *14*, 305–308.
15. I. Hua, R. H. Hoechemer, M. R. Hoffmann, *J. Phys. Chem.* **1995**, *99*, 2335–2342.
16. T. Lepoint, F. Mullie, *Ultrason. Sonochem.* **1994**, *1*, S13–S22.
17. S. I. Nikitenko, *Adv. Phys. Chem.* **2014**, Article ID 173878.
18. S. I. Nikitenko, R. Pflieger, *Ultrason. Sonochem.* 2016, doi:10.1016/j.ultsonch.2016.02.003.
19. T. J. Mason, J. P. Lorimer, *Sonochemistry: Theory, Applications and Uses of Ultrasound in Chemistry*, Ellis Horwood Publishers, Chichester, 1988, 252.
20. F. M. Nowak, *Sonochemistry: Theory, Reactions and Syntheses, and Applications (Chemistry Engineering Methods and Technology)*, Nova Science Publishers Inc., USA, 2010.
21. T. Y. Wu, N. Guo, C. Y. Teh, J. X. W. Hay, *Advances in Ultrasound Technology for Environmental Remediation*, SpringerBriefs in Green Chemistry for Sustainability, 2013.
22. A. Henglein, *Advances in Sonochemistry*, JAI Press, London, 1993.
23. A. Kotronarou, G. Mills, M. R. Hoffmann, *J. Phys. Chem.* **1991**, *95*, 3630–3638.
24. H. Z. Heusinger, *Lebensm. Unters. Forsch.* **1987**, *185*, 447–456.
25. E. Fuchs, H. Z. Heusinger, *Lebensm. Unters. Forsch.* **1994**, *198*, 486–490.
26. E. B. Flint, K. S. Suslick, *Science* **1991**, *253*, 1397–1399.
27. M. A. Margulis, *Sonochemistry and Cavitation*, Gordon and Breach Publishers, Amsterdam, 1995, 540.

28. P. A. Tatake, A. B. Pandit, *Chem. Eng. Sci.* **2002**, *57*, 4987–4995.
29. A. V. Prabhu, P. R. Gogate, A. B. Pandit, *Chem. Eng. Sci.* **2004**, *59*, 4991–4998.
30. A. Brotchie, F. Grieser, M. Ashok kumar, *J. Phys. Chem. C* **2008**, *112*, 10247–10250.
31. D. G. Wayment, D. J. Casadonte Jr., *Ultrason. Sonochem.* **2002**, *9*, 189–195.
32. M. Beckett, I. Hua, *J. Phys. Chem. A* **2001**, *105*, 189–195.
33. H. Ferkous, O. Hamdaoui, S. Merouani, *Ultrason. Sonochem.* **2015**, *26*, 40–47.
34. F. G. Blake, Onset of cavitation in liquids, In: *Acoustics Research Laboratory*, Harvard University, Cambridge, 1949.
35. S. de La Rochebrochard, J. Suptil, J.-F. Blais, E. Naffrechoux, *Ultrason. Sonochem.* **2012**, *19*, 280–285.
36. M. Lim, M. Ashokkumar, Y. Son, *Ultrason. Sonochem.* **2014**, *21*, 1988–1993.
37. Y. Kojima, Y. Asakura, G. Sugiyama, S. Koda, *Ultrason. Sonochem.* **2010**, *17*, 978–984.
38. T. J. Mason, *Practical Sonochemistry: User's Guide to Applications in Chemistry and Chemical Engineering*, Ellis Horwood Ltd, New York, 1992, 150.
39. A. Henglein, M. Gutierrez, *J. Phys. Chem.* **1993**, *97*, 158–162.
40. H. Delmas, N. Tuan Le, L. Barthe, C. Julcour-Lebigue, *Ultrason. Sonochem.* **2015**, *15*, 51–59.
41. E. Luengo, S. Condón-Abanto, S. Condón, I. Álvarez, J. Raso, *Sep. Purif. Technol.* **2014**, *136*, 130–136.
42. N. T. Le, C. Julcour-Lebigue, H. Delmas, *Ultrason. Sonochem.* **2013**, *20*, 1203–1210.
43. T. J. Mason, *Sonochemistry*, Chemistry Primers, Oxford, 2000, 68.
44. M. H. Entezari, P. Kruus, *Ultrason. Sonochem.* **1996**, *3*, 19–24.
45. A. E. Kandjani, M. F. Tabriz, B. Pourabbas, *Mater. Res. Bull.* **2008**, *43*, 645–654.
46. K. S. Suslick, J. J. Gawienowski, P. F. Schubert, H. H. Wang, *Ultrasonics* **1984**, 33–36.
47. K. Weninger, R. Hiller, B. P. Barber, D. Lacoste, S. J. Putterman, *J. Phys. Chem.* **1995**, *99*, 14195–14197.
48. K. Gaddam, M. Cheung, *Ultrason. Sonochem.* **2001**, *8*, 103–109.
49. J. Rooze, E. V. Rebrov, J. C. Schouten, J. T. F. Keurentjes, *Ultrason. Sonochem.* **2013**, *20*, 1–11.
50. T. J. Mason, J. P. Lorimer, *Sonochemistry: Theory, Applications and Uses of Ultrasound in Chemistry*, Wiley-Interscience, New York, 1989, 580.
51. S. I. Nikitenko, C. Le Naour, P. Moisy, *Ultrason. Sonochem.* **2007**, *14*, 330–336.
52. S. de La Rochebrochard d'Auzay, J.-F. Blais, E. Naffrechoux, *Ultrason. Sonochem.* **2010**, *17*, 547–554.

Chapter 3

Ultrasonic Parameters Estimation

Many parameters can change the effects of ultrasound on a chemical reaction, a material preparation or another process. It is also sometimes difficult to replicate a specific experiment described in the literature because the equipment is very different from one research group to another. In addition, some publications do not specifically describe the characteristics of the studied sonochemical system. For these reasons, this chapter is essential for you in order to adopt good practices when you use ultrasound. It presents how to improve the reproducibility of an experiment even if it is performed with another type of sonochemical equipment.

For example in organic chemistry, all the experimental data needed to reproduce a reaction are usually clearly indicated (solvent, reaction temperature, agitation mode, reaction time, etc.). It must be the same when ultrasound is used: experimental conditions and sonochemical parameters have to be reported.

1. Frequency

The frequency, expressed in kHz or MHz, is a key parameter to report when we use ultrasound. As explained in the previous chapter, **low frequencies (20–80 kHz)** preferentially lead to physical effects (shockwaves, microjets, microconvection, etc.) whereas **high frequencies (150–2,000 kHz)** favor the chemical effects (radical production, etc.). In addition, a difference of frequencies (even within each frequency area) could dramatically change the sonochemical effects. Generally, the frequency is a fixed characteristic of ultrasonic equipment since the reactor or the probe are directly designed and optimized according to the frequency. This data, provided by the constructor, has to be systematically reported.

Multifrequency systems could also be used.[1,2] In this case, the imposed frequency has to be clearly reported.

2. Electric Power

The "nominal electric power" or the "electric power input" is the power delivered by the generator. In fact, it is essential to estimate the electric power converted to acoustic power in the medium (*via* the transducer, see Chapter 4, page 41). The most economical and practical method is to connect the generator to a wattmeter. The electric power P_{elec} of the used generator was calculated by the difference between total consumed power P_{total} by the generator and its power of standby P_0 (Eq. (1)).

$$P_{elec} = P_{total \ (measurement)} - P_0.$$
(1)

As shown in Figure 1, P_{total} could represent an interesting data to consider the energy balance in the process in its entirety, but P_{elec} only corresponds to the power provided by the generator and converted into acoustic power delivered in the medium.

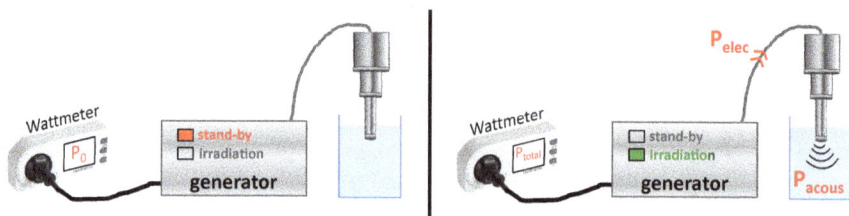

Figure 1: Determination of the electric power (P_{elec}) by measuring the power of standby (P_0) and the total consumed power (P_{total}).

⚠ It is important to have in mind that the measurement performed with the wattmeter is an indicative data on the overall energy consumption of the process, but an important part of this energy could be only due to the intrinsic consumption of the equipment (not really optimized for extrapolation to larger scale processes).

3. Acoustic Power

Often even if it is the only data reported concerning the sonochemical parameters in the literature, the electric power is not sufficient to describe an ultrasonic system. Indeed, huge differences could be found between the electrical power consumed (P_{total}) by the equipment and the acoustic power (P_{acous}), the power really dissipated in the irradiated fluid. Thus, the estimation of the acoustic power by calorimetric measurements is considered as a representative and practical method.

Above the cavitation threshold, a part of the acoustic energy is converted into heat by absorption. If we know the mass m (expressed in g) of the irradiated fluid and its specific heat capacity c_p (in $J \cdot g^{-1} \cdot K^{-1}$), the initial temperature rise per unit of time (($dT/dt)_{t0}$ expressed in $K \cdot s^{-1}$) induced by the ultrasound can be easily converted into energy input, using Eq. (2):

$$P_{acous} = m \times c_p \times \left(\frac{dT}{dt}\right)_{t_0}. \tag{2}$$

The thermal capacitance of the reactor was neglected in relation to the liquid. In addition, the sonicated medium was assumed to be perfectly mixed, especially at low frequencies.[3,4]

Generally, the acoustic power is reported by unit of volume as a volume acoustic power $P_{acous.vol}$, expressed in $W \cdot L^{-1}$. If the irradiated fluid is complex (mixture) or if the heat capacity is not available, it could be appropriate to estimate the acoustic power in water.

In 1991, Margulis *et al.* described a very accurate measurement of the power absorbed in a system, comparing the ultrasonic conditions to the Joule heat in the case of a calibrated thermistor.[5] Many methods to determine precisely the acoustic power by calorimetry were now published.[6–8] In addition, other physical measurements, less commonly applied, are based on the use of hydrophones,[9,10] water-immersed radiation pressure balances (radiometric methods)[11] or metal foils (determination of the erosion).[12]

A rapid method to estimate the acoustic power by calorimetry:

In order to estimate P_{acous} in water, you first have to know the mass of water that will be irradiated (m, in grams). To determine $(dT/dt)_{t_0}$, measure the temperature rise of the medium as a function of the time (for example during 5 min or 10 min, each 30 s). Continue to observe the temperature evolution after stopping irradiation: this decrease corresponds to the used non-adiabatic system and could be a correction factor of the increase of the temperature under ultrasound. $(dT/dt)_{t_0}$ is the sum of the slope of rise during sonication and slope of decrease after stopping irradiation. It is expressed in $K \cdot s^{-1}$. Then, calculate P_{acous} according the Eq. (2) with c_p (water) = $4.18\ J \cdot g^{-1} \cdot K^{-1}$ at 20°C.

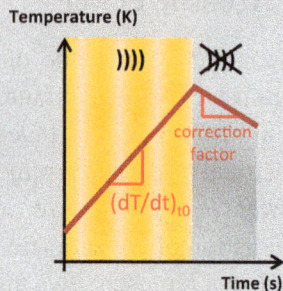

Interestingly, from calorimetric measurements, two other parameters can be determined: (i) the ultrasonic intensity I_{US} (Section 3 of this chapter); (ii) the acoustic efficiency $E_{acous/elec}$ which is the ratio between acoustic power and electric power. This latter determines the conversion of the electric energy into acoustic energy delivered in the medium. It depends on the equipment and the medium.

4. Ultrasonic Intensity

In the case of a planar or spherical wave, the sound pressure (P_A, expressed in Pascal) and the theoretical ultrasonic intensity (I_{max}, expressed in $W \cdot m^{-2}$) are related by Eq. (3):

$$I_{max} = \frac{P_A^2}{2\rho v},$$ (3)

where ρ is the density of the irradiated fluid ($kg \cdot m^{-3}$) and v is the speed of the ultrasonic wave ($m \cdot s^{-1}$).

The term ρv represents the acoustic impedance of the medium, whose the value is $1.5 \cdot 10^6\ kg \cdot m^{-2} \cdot s^{-1}$ for water.[13] The determination of the experimental ultrasonic intensity (I_{US}, expressed in $W \cdot cm^{-2}$) is estimated by measuring the acoustic power per unit area of the probe (Eq. (4)):

$$I_{US} = \frac{P_{acous}}{S_{probe}},$$

(4)

where P_{acous} is the acoustic power (W) and S_{probe} is the surface of the irradiating probe (cm^{-2}).

5. Radical Production

The most conventional chemical dosimeter used to characterize the production of free radicals obtained by ultrasonic irradiation is certainly potassium iodide (KI). The reaction yield to produce iodine by ultrasonic irradiation of an aqueous solution of potassium iodide in a given time, known as the "Weissler reaction", is considered as a relative measure of acoustic cavitation performance.[12] Indeed, the transient cavitation leads to the decomposition of water vapor into $HO^\bullet$ and $H^\bullet$ radicals (Reaction 1). Iodide ions are then oxidized by radicals formed $HO^\bullet$ and transferred in liquid phase (Reaction 2). In the absence of radical scavenger, hydroxyl radicals form hydrogen peroxide by recombination in gas phase or at the liquid/gas interface (Reaction 3).

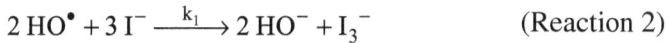

$$H_2O \xrightarrow{\;))))\;} HO^\bullet + H^\bullet \qquad \text{(Reaction 1)}$$

$$2\,HO^\bullet + 3\,I^- \xrightarrow{\;k_1\;} 2\,HO^- + I_3^- \qquad \text{(Reaction 2)}$$

$$HO^\bullet + HO^\bullet \xrightarrow{\;k_2\;} H_2O_2. \qquad \text{(Reaction 3)}$$

The rate constants of Reactions 2 and 3 are comparable and very high: $k_1 = 1.1.10^{10}\ L \cdot mol^{-1} \cdot s^{-1}$ and $k_2 = 6.2.10^9\ L \cdot mol^{-1} \cdot s^{-1}$. The advantage of this method is the easiness to determine I_3^- ions concentration by UV–Visible spectrophotometry at the wavelength of 355 nm ($\varepsilon_{355nm} = 26,303\ L \cdot mol^{-1} \cdot cm^{-1}$) and to deduct the concentration of $HO^\bullet$ radicals.[14,15] Some other methods can also be used to quantify the produced radical species under ultrasound such as terephthalate dosimetry (from

---

A rapid method to estimate the radicals production in water[16,17]:

Example of a 10 mL volume irradiated at 20 kHz by an ultrasonic horn. Each experiment is maintained at 25°C using a cooling system and is triplicated under the following conditions.

In this example, the potassium iodide dosimetry is carried out from a 0.1 mol·L^{-1} KI solution during 1 hour. The [I_3^-] concentration is monitored by a UV–Visible spectrophotometer at a wavelength of 355 nm using the Beer–Lambert law. From these data, it is easy to determine the formation rate of I_3^- under ultrasonic conditions (mol·L^{-1}) and then estimate the HO$^{\bullet}$ production.

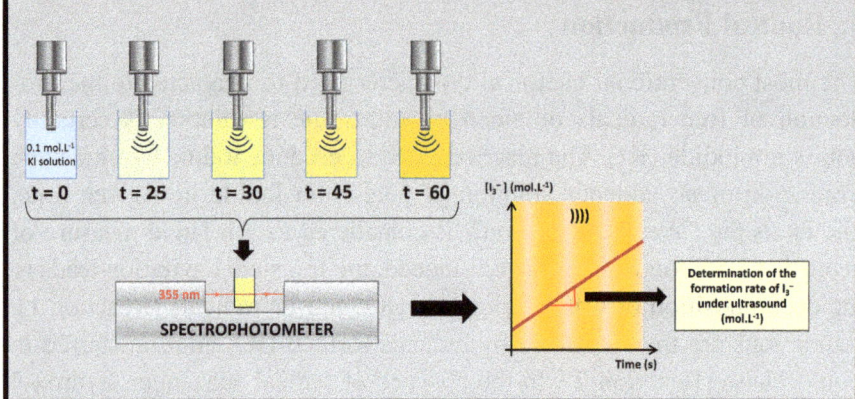

---

terephthalic acid in alkaline solution, generation of highly fluorescent 2-hydroxyterephthalate ions),[18] Fricke dosimetry (oxidation of Fe^{2+}; determination of Fe^{3+} formation by photometry),[19,20] nitrite and nitrate dosimetry and pH evolution (from dissolved nitrogen and oxygen gas under sonication in the presence of Na_2CO_3 and $NaHCO_3$; analysis of NO_2^- and NO_3^- using ion chromatography).[21]

Another very accurate method, but not available in all laboratories, is the spin trapping (using diamagnetic compounds) of radicals formed under ultrasound and their monitoring by Electron Paramagnetic Resonance (EPR) spectroscopy.[22–24]

In non-aqueous media, other radicals can be generated other than HO$^{\bullet}$. The sonolysis of different organic solvents was investigated by EPR in low-energy ultrasonic baths.[25,26] The sonolysis is more important for liquids with low volatility. All the solvents are not inert under ultrasonic conditions. For example, using dimethylformamide (DMF) as solvent in sonochemical reactions may be problematic due to the cleavage of the C–N bond under ultrasonic conditions.

6. Sonochemical Efficiency

Based on the estimations of electric/acoustic powers and the rate of anion formation determined by dosimetry (v_{ion}, expressed in $mol \cdot s^{-1}$), it is interesting to calculate the sonochemical efficiency (SE) ($mol \cdot J^{-1}$), detailing SE_{elec} taken into account the electric power (Eq. (5)) and SE_{acous} for acoustic power (Eq. (6)).

$$SE_{elec} = \frac{n_{ion}}{E_{elec}} = \frac{v_{ion}}{P_{elec}}, \tag{5}$$

$$SE_{acous} = \frac{n_{ion}}{E_{acous}} = \frac{v_{ion}}{P_{acous}}, \tag{6}$$

where n_{ion} is the mole number of the considered ion (I_3^-, NO_2^-, NO_3^-, etc., expressed in mole), E_{elec} is the electric energy (J), v_{ion} is the formation rate of the considered ion ($mol \cdot s^{-1}$), E_{acous} is the acoustic energy (J) and P_{acous} is the acoustic power (W).

Interestingly, SEs give the overall efficiency of the ultrasonic system taking into account both the power loss and the formation of radicals, constituting an efficiency assessment method to compare different ultrasonic conditions/reactors.[27,28]

7. Speed of Ultrasonic Wave

The propagation speed of the ultrasonic wave is 1,435 $m \cdot s^{-1}$ in deionized water at 25°C. It can be calculated in pure fluids as a function of the temperature and the number of carbons according to the following equation[29]:

$$v = a + \frac{b}{C_n} - \left[d + \frac{e}{C_n} \right] T, \tag{7}$$

where v is the speed of ultrasonic wave $m \cdot s^{-1}$, T the temperature (°C), C_n the number of carbon atoms, a, b, d, and e some constants.

The increase of the speed is inversely proportional to the number of carbon atoms. The constants a, b, d, and e were evaluated for linear alkanes and terminal alkenes.[30] Many studies were used to estimate the speed of the ultrasonic wave in different liquids and especially in mixtures of organic solvents.[31]

⚠ It is not necessary to determine and provide this parameter in your work, but it could become very useful if your studies are about the influence of the liquid medium on the sonochemical effects.

8. Cavitation Regions

Because of the intrinsic characteristics of the ultrasonic waves, it is impossible to obtain a homogeneous phenomenon over the entire volume of a liquid. The ultrasonic wave propagates from the transmission source in a cone defined by Eq. (8). 90% of the energy propagates in this volume; therefore, the geometry of the reactor is of major importance in ultrasonic irradiation.

$$\sin \theta = \frac{1{,}2\lambda}{d}, \tag{8}$$

where θ is the half of the cone angle (°), λ is the wavelength (m) and d, the diameter of the emission source (m).

Different methods are used to determine the "active" zone in a sonoreactor: optical (light scattering), chemical, electrical and mechanical methods,[32] measurement of the local temperature increase (silicone thermocouple), the use of microdiffusion sensors, the use of thermal microvibration sensors[33] or sonoluminescence (SL).

9. Sonoluminescence

The extreme conditions created by the collapse of the bubble have also consequences such as the emission of photons. This phenomenon, called SL, is intimately linked to cavitation, but its physical origin is still poorly understood.

The SL is often observed due to the addition of luminol (3-aminophtalhydrazide) which oxidizes in the presence of HO• radicals into

---

Mechanical determination of the active zone in an ultrasonic bath[32]:

In order to determine the "active" zone in a reactor or in an ultrasonic bath (especially for low frequencies ultrasonic irradiations), it is possible to dive vertically an aluminum foil in the irradiated liquid. The physical effects due to the implosion of cavitation bubbles lead to important impacts on the aluminum foil. This method is a rapid way to identify areas where the ultrasound is most effective.

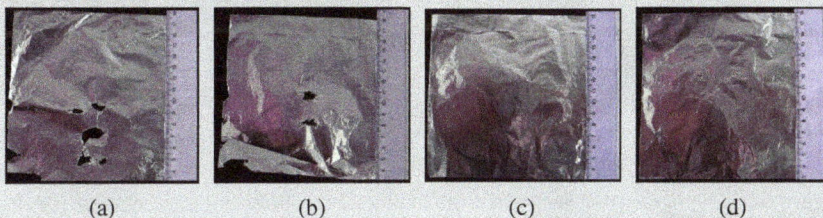

(a)　　　　　　　(b)　　　　　　　(c)　　　　　　　(d)

　　These four images* show the result of this test with an aluminum foil under various conditions after 5 min of ultrasonic irradiation in an ultrasonic bath: (a) 35 kHz, 100% of the maximal power; (b) 35 kHz, 50% of the maximal power; (c) 130 kHz, 100% of the maximal power; (d) 130 kHz, 50% of the maximal. This example shows that the physical effects depend on the frequency and power of ultrasound.

*Pictures reprinted from Ref. 32, Copyright (2016), with permission from John Wiley & Sons, Inc.

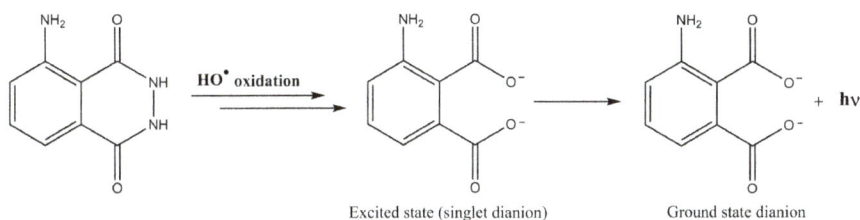

Excited state (singlet dianion)　　　　　Ground state dianion

Scheme 1:　Chimiluminescence of luminol.

3-aminophthalic acid, presenting electrons in an excited state (Scheme 1). The deenergization of these electrons causes visible blue light emission of energy consisting in photons of wavelength $\lambda = 430$ nm. Generally, this method is not quantitative but allows a precise mapping of effective zones of a sonochemical reactor. In reality, this method corresponds to chemiluminescence (Figure 2).[34]

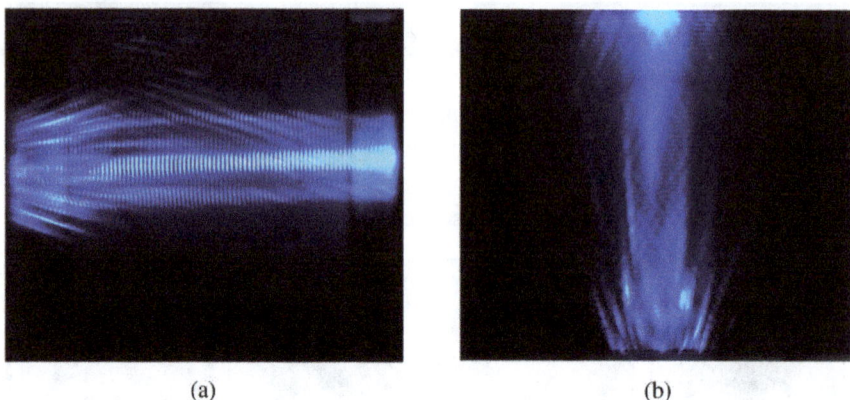

(a)   (b)

Figure 2: Photographs of sonochemical luminescence from luminol solution (electric power of 60 W): (a) 472 kHz (ultrasonic emission on the side), (b) 422 kHz (ultrasonic emission on the bottom). Reprinted with permission from Ref. 34. Copyright (2016). Elsevier.

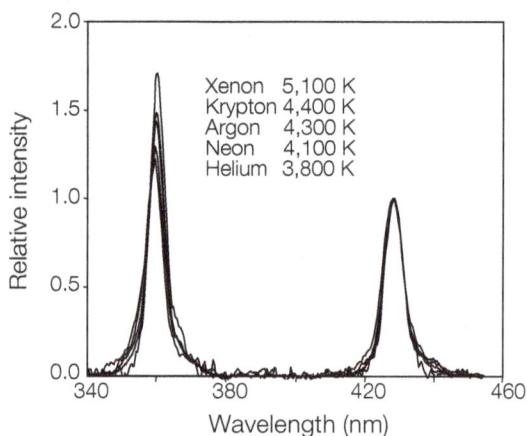

Figure 3: The multibubble SL spectra obtained from solutions of $Cr(CO)_6$ in octanol with different dissolved gases (Xe, Kr, Ar, Ne and He) with respective emission temperatures of 5,100 K, 4,400 K, 4,300 K, 4,100 K, and 3,800 K. Reprinted with permission from Macmillan Publishers Ltd: from Ref. 35. Copyright (2016).

Using specific detectors, UV–Visible spectra of the SL (Figure 3) observed in the sonochemical reactor allow examining the activity of radical or excited species formed in the gas phase of the bubble (HO$^\bullet$, H$^\bullet$, X$^\bullet$,

OH*, C_2*, CN*, etc.) during the implosion and their possible release in the liquid phase.[34,35]

10. Conclusions

To resume, some parameters are essential to report when you use ultrasound in chemistry. The **frequency**, the **acoustic power** and the **details on the equipment** (dimensions of the ultrasonic bath or probe, geometry of the reactor, volume of solvent) have to be systematically determined and specified. The **experimental conditions** are also important as for all experiments in chemistry: solvent, temperature, pressure, reaction time, etc.

Some other important sonochemical parameters should be indicated or estimated to better understand the involved mechanisms. The power used by the generator (P_{total}) could be interesting to have an idea on the total consumption of the process (the input necessary energy). By determining P_{elec}, it is easy to provide the acoustic efficiency $E_{acous/elec}$ giving essential information on the efficiency of the used equipment. The ultrasonic intensity I_{US} can also provide interesting data on the equipment.

The estimation of radical production (for example, by calorimetric methods) can become essential to demonstrate a radical mechanism. From these data, it will be easy to calculate the sonochemical efficiencies (SE_{elec} and SE_{acous}) that are very representative information on the ultrasonic equipment.

References

1. R. Feng, Y. Zhao, C. Zhu, T. J. Mason, *Ultrason. Sonochem.* **2002**, *9*, 231–236.
2. T. J. Mason, A. J. Cobley, J. E. Graves, D. Morgan, *Ultrason. Sonochem.* **2011**, *18*, 226–230.
3. K. Shinobu, T. Kimura, T. Sakamoto, T. Kondo, H. Mitome, *Ultrason. Sonochem.* **2003**, *10*, 149–156.
4. T. J. Mason, J. P. Lorimer, D. M. Bates, Y. Zhao, *Ultrason. Sonochem.* **1994**, *1*, 91–95.
5. M. A. Margulis, N. A. Maximenko, *Advances in Sonochemistry*, JAI Press, London, 1991, 253–292.

6. M. A. Margulis, I. M. Margulis, *Ultrason. Sonochem.* **2003**, *10*, 343–345.
7. Ratoarinoro, F. Contamine, A. M. Wilhelm, J. Berlan, H. Delmas, *Ultrason. Sonochem.* **1995**, *10*, 43–47.
8. T. Uchida, T. Kikuchi, M. Yoshioka, Y. Matsuda, R. Horiuchi, *Acoust. Sci. Technol.* **2015**, *36*, 445–448.
9. G. R. Harris, *Ultrasound. Med. Biol.* **1995**, *11*, 803–817.
10. Y. Zhou, L. Zhai, R. Simmons, P. Zhong, *J. Acoust. Soc. Am.* **2006**, *120*, 676–685.
11. P. N. T. Wells, M. A. Bullen, D. H. Follet, H. F. Freundlich, J. A. James, *Ultrasonics* **1963**, *1*, 106–110.
12. A. Weissler, *J. Am. Chem. Soc.* **1959**, *81*, 1077–1081.
13. J.-M. Löning, C. Horst, U. Hoffmann, *Ultrason. Sonochem.* **2002**, *9*, 169–179.
14. S. Koda, T. Kimura, T. Kondo, H. Mitome, *Ultrason. Sonochem.* **2003**, *10*, 149–156.
15. Y. Kojima, Y. Asakura, G. Sugiyama, S. Koda, *Ultrason. Sonochem.* **2010**, *17*, 978–984.
16. M. H. Entezari, P. Kruus, *Ultrason. Sonochem.* **1994**, *1*, S75–S80.
17. S. de La Rochebrochard, J. Suptil, J.-F. Blais, E. Naffrechoux, *Ultrason. Sonochem.* **2012**, *19*, 280–285.
18. G. Mark, A. Tauber, R. Laupert, H.-P. Schuchmann, D. Schulz, A. Mues, C. von Sonntag, *Ultrason. Sonochem.* **1998**, *5*, 41–52.
19. Y. Iida, K. Yasui, T. Tuziuti, M. Sivakumar, *Microchem. J.* **2005**, *80*, 159–164.
20. S. Merouani, O. Hamdaoui, F. Saoudi, M. Chiha, *J. Hazard. Mater.* **2010**, *178*, 1007–1014.
21. S. de La Rocherbrochard d'Auzay, J.-F. Blais, E. Naffrechoux, *Ultrason. Sonochem.* **2010**, *17*, 547–554.
22. Q.-A. Zhang, Y. Shen, X.-H. Fan, J. F. García Martín, X. Wang, Y. Song, *Ultrason. Sonochem.* **2015**, *27*, 96–101.
23. K. Makino, M. M. Mossoba, P. Riesz, *J. Am. Chem. Soc.* **1982**, *104*, 3537–3539.
24. P. Riesz, D. Berdahl, C. L. Christman, *Environ. Health Perspect.* **1985**, *64*, 233–252.
25. J. P. Lorimer, D. Kershaw, T. J. Mason, *J. Chem. Soc. Faraday Trans.* **1995**, *91*, 1067–1074.
26. B. A. Niemczewski, *Ultrasonics* **1980**, *18*, 107–110.
27. P. R. Gogate, I. Z. Shirgaonkar, M. Sivakumar, P. Senthilkumar, N. P. Vicharen A. B. Pandit, *AIChE J.* **2001**, *47*, 2526–2538.
28. G. Chatel, L. Leclerc, E. Naffrechoux, C. Bas, N. Kardos, C. Goux-Henry, B. Andrioletti, M. Draye, *J. Chem. Eng. Data* **2012**, *57*, 3385–3390.

29. J.-P. Bazureau, M. Draye, *Ultrasound and Microwaves: Recent Advances in Organic Chemistry*, Transworld Research Network, Kerala, 2011, 241.

30. Z. Wang, A. Nur, *J. Acous. Soc. Am.* **1991**, *89*, 2725–2730.

31. M. Hasan, D. F. Shirude, A. P. Hiray, U. O. Kadam, A. B. Sawant, *J. Mol. Liq.* **2007**, *135*, 32–37.

32. H. M. Santos, C. Lodeiro, J.-L. Capelo-Martinez, *The Power of Ultrasound in Ultrasound in Chemistry: Analytical Applications*, Wiley-VCH, Weinheim, 2009, 8–9.

33. B. Pugin, A. T. Turner, *Advances in Sonochemistry*, JAI Press, London, 1990, 81–118.

34. K. Yasuda, T. Torii, K. Yasui, Y. Iiad, T. Tuziuti, M. Nakamura, Y. Asakura, *Ultrason. Sonochem.* **2007**, *14*, 699–704.

35. W. B. McNamara, Y. Didenko, K. S. Suslick, *Nature* **1999**, *401*, 772–775.

Chapter 4

Ultrasonic Equipment

This chapter will give you information on the existing equipment used to develop sonochemical processes in labs. There are different devices from the most mundane to the most advanced that produce ultrasound. The following sections present each of these systems by highlighting their practical advantages and disadvantages.

1. Transducers

A device generating ultrasound is commonly called a **transducer**. The technology of transducers is based on the properties of piezoelectric materials to convert electrical energy into mechanical energy. This mechanical vibration is then transmitted into the liquid medium in the form of ultrasonic wave.

Piezoelectric transducers utilize the reverse piezoelectric effect of natural or synthetic single crystals (such as quartz) or ceramics dispersed with barium titanate ($BaTiO_3$), lead metaniobate ($PbNb_2O_6$), or lead zirconate titanate (LZT with the chemical formula $Pb(Zr_xTi_{1-x})O_3$) that can be easily machined. They occur most often in the form of a disc, a plate or a ring on the faces of which are two fixed metalized electrodes. When an electric voltage is applied to these two electrodes, the material expands or compresses according to the orientation of the voltage with respect to the polarization of the ceramic. Displacement amplitudes (or vibrations) of the ceramics are very low (about few micrometers), but the emitting face of a transducer is at the maximum point of vibration. The transducer, consisting of the assembly, develops a maximum displacement to particular frequencies which depend on its geometry: these frequencies are called **resonance frequencies**.[1]

To create an ultrasonic wave, an electrical voltage is applied to the transducer with a frequency equal to its resonance frequency. An electric generator is to be used to transform the mains voltage (220 V — 50/60 Hz) to an AC voltage with the resonance frequency of the system (for example, 1,000 V — 20 kHz). The conversion efficiency of electrical energy provided by the generator into the acoustic energy transmitted to the medium are currently about **30–40%** in the best cases for low frequency ultrasound. Their optimization represents a major technological challenge for the development at an industrial scale of the potential applications using ultrasound.

Conventionally, the ceramic is bonded or clamped to a mass (and sometimes against mass) of metal or glass in order to isolate the liquid medium in which is transmitted ultrasound (Figure 1). The reaction vessel is typically immersed in the coupling fluid contained in the bath (**indirect irradiation mode**). However, the bath itself can be used as the reaction vessel (**direct irradiation mode**). Titanium or titanium alloy is often used as a waveguide (probe). In the case of 20 kHz irradiation, the titanium probe is generally 12.7 cm length (half wavelength) or a multiple of this number (velocity of sound in titanium is about 5,080 m·s^{-1}). In this case, the ultrasonic irradiation is direct in the studied fluid.

Generally, piezoelectric devices must be cooled if they are to be used for long periods at high temperatures because the ceramic material will degrade under these conditions.

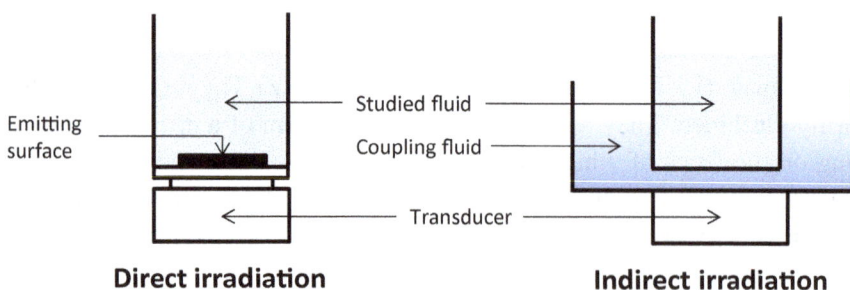

Emitting surface — Studied fluid — Coupling fluid — Transducer

Direct irradiation   **Indirect irradiation**

Figure 1: Example of direct and indirect irradiation modes.

Not only piezoelectric transducers...[2]

Piezoelectric transducers are the most widely used for sonochemical application. However, some other types of ultrasound emissions are possible, for example from **magnetostrictive transducers** (ferromagnetic materials placed under an alternating magnetic field) or **electrostrictive transducers** (ceramic expand and contract in an alternating electric field).

Piezoelectric transducers	Magnetostrictive transducers
⇨ relatively inexpensive;	⇨ more expensive than piezoelectric for similar power ratings;
⇨ relatively small and light;	⇨ heavier and bulkier than piezoelectric;
⇨ damaged at temperature >150°C;	⇨ can be operated at temperatures >250°C with specific precautions;
⇨ will age considerably (reduced power input with high temperatures or long irradiation times);	⇨ will not degrade or fail over time by their very nature;
⇨ may be damaged by large impact;	⇨ extremely resistant to mechanical damage, such as large impacts;
⇨ structure will be damage if operated "dry".	⇨ no damage when operated "dry".

As is evident from the table above, piezoelectric transducers are normally used with small-volume processes (at lab). When large volumes and/or long, continuous reaction times are required, the more robust magnetostrictive transducer may be the preferred option.

2. Ultrasonic Cleaning Baths

Ultrasonic cleaning baths (Figure 2) are the cheapest and the most widely used ultrasound emitters in laboratories, where they are used to create or break emulsions, dissolve compounds, degas an elution solvent, clean glassware, etc.

This type of apparatus generates frequencies between 20 kHz and 60 kHz at low acoustic intensities, generally between 1 $W \cdot cm^{-2}$ and 5 $W \cdot cm^{-2}$ in order to not damage the tank during the acoustic cavitation. The vibration

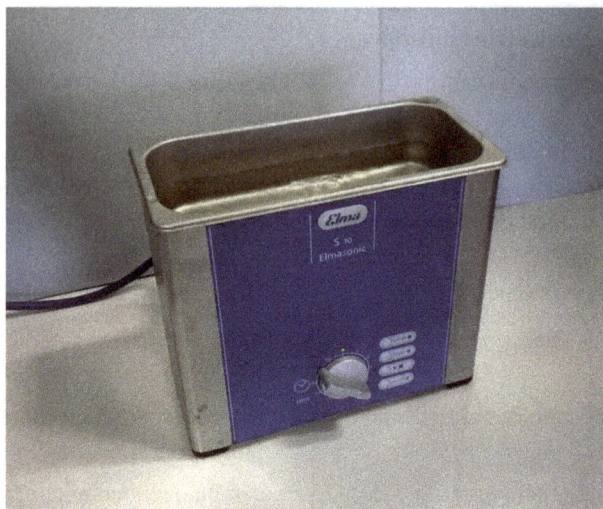

Figure 2: Picture of an ultrasonic bath using at lab.

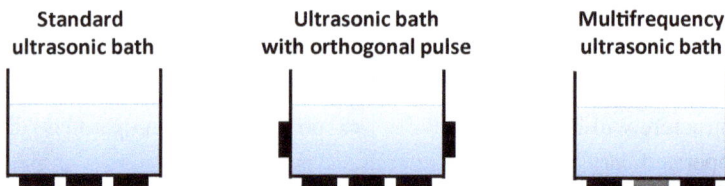

Figure 3: Examples of ultrasonic baths showing different irradiation types.

source is usually placed at the bottom of a tank although some alternatives exist (Figure 3).

The ultrasonic irradiation using bath is **indirect**. A liquid, typically water, is used to transmit power from the irradiation zone to the reactor where the chemical reaction occurs.[3] Therefore, the ultrasonic field is not homogeneous throughout the entire volume of water. The location of the reactor in the tank is essential to use the maximum acoustic intensity in the reaction involved. Moreover, if the bath temperature is not controlled, it can heat under the effect of ultrasound and make some variation of the temperature of the reaction mixture over time. These devices therefore have some limitations in terms of reproducibility of experiments.[4]

One of the major shortcomings of the ultrasound baths is the directional sensitivity of the ultrasound waves in the bath, which creates a non-homogeneous energy dissipation pattern.[5]

3. Ultrasonic Probes

The term "ultrasonic probe" is often defined, by abuse of language, a complete irradiation system, but in fact, it is the final part of the system composed of (i) the generator, (ii) the ultrasonic transducers, (c) the amplification probe (booster) or an against mass, and (d) the ultrasonic probe or detachable probe (Figure 4). The generator transforms the usual electric current into high frequency electrical energy. The converter transforms the electrical energy into mechanical energy at a given frequency. The amplifier probe increases the amplitude of the ultrasonic wave while the ultrasonic wave probe transmits the reaction medium.

These probes have two fundamental differences with ultrasonic baths: (1) they allow **direct irradiation** of the medium; (2) they have power 100 times higher than those provided by baths. They are generally made of a titanium alloy in order to resist erosion induced by cavitation (Figure 5).

The length of the probe is exactly equal to a multiple of half the wavelength (e.g., 12.5 cm at 20 kHz). The generated intensity is inversely proportional to the area of the radiating area and thereof is selected according to the volume to be treated. When using such high-energy devices, a steady increase of the medium temperature is observed. Importantly, it can lead to the modification of the physico-chemical properties of the medium and even at the boiling of the more volatile compound(s), such as the solvent. The use of a jacketed reactor, with a circulating cooling fluid, allows regulation of the reaction temperature of the medium.

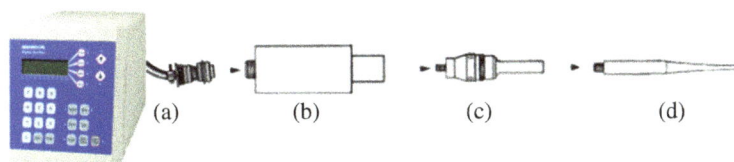

Figure 4: Direct ultrasonic irradiation system composed of (a) a generator, (b) a transducer, (c) a booster and (d) a detachable probe.

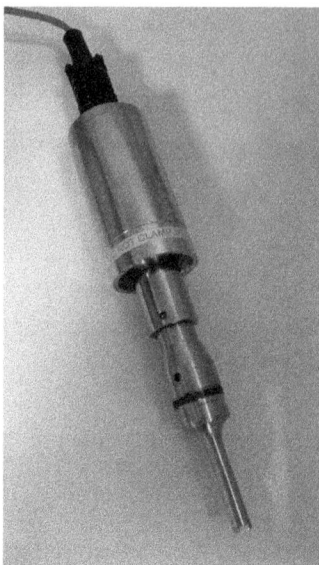

Figure 5: Picture of an ultrasonic probe system (transducer, amplification probe and ultrasonic probe).

The **pulsed mode** can be used to allocate the same amount of energy in different time intervals and to adapt the droplet size distributions for emulsification processes, for example.

The commonly used ultrasonic horn generates localized cavitation below its converging tip resulting in a dense bubble cloud near the tip and limiting diffusion of reactive components into the bubble cloud or reactive radicals out of the bubble cloud. To improve contact between reactive components, some researchers designed novel ultrasonic probes.[6] For example, Figure 6 reports the difference in terms of acoustic field distribution and sonoluminescence imaging for two different shapes of probe at 20 kHz.

4. Reactors

4.1. Cup-horn reactors

The "cup-horn" reactors (Figure 7) allow **direct and intense irradiation** of the medium in which they are located. This type of device, available from many suppliers, is comparable to an ultrasonic bath with high intensity.

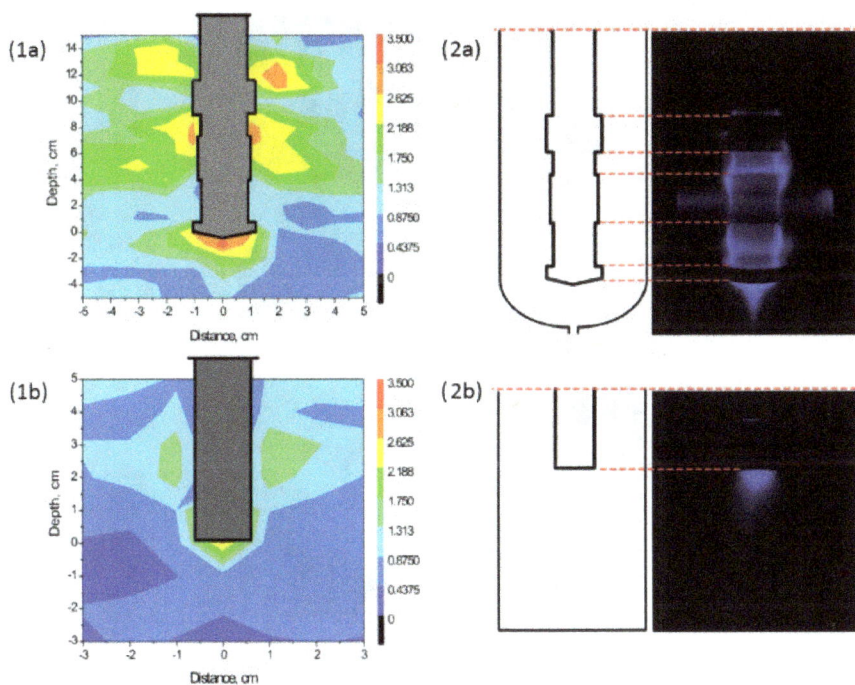

Figure 6: (1) Acoustic field distribution determined from hydrophone measurements (color key in volts, initial temperature of 20°C); (2) Sonoluminescence imaging in the reactor (initial luminol concentration of 10^{-3} mol·L^{-1}, pH of 11.3 and initial temperature of 20°C) for (a) designed probe and (b) standard probe. Reprinted with permission from Ref. 6. Copyright (2016) Elsevier.

However, the distribution of the ultrasonic field can be much more uniform if the proper geometry of the reactor is used. Thus, the cup-horn reactors allow irradiation upwardly, generally 50 times more intense than a traditional ultrasonic bath.[7]

Figure 8 shows a cup-horn reactor constituting of a piezoelectric ceramic set at the bottom of a jacketed chemical reactor. The ceramics are generally protected by borosilicate glass window with a thickness adjusted to a factor of the quarter of the wavelength λ to avoid reflection phenomena.[7] When the volumes to be treated are low, a glass container (tube, flask, beaker, etc.) can be dipped into the reactor. In this configuration, the ultrasonic irradiation becomes indirect and the energy supplied to the reaction medium is lower.

Figure 7: 20-kHz Cup-horn reactor with glass double jacket to cool the solution.

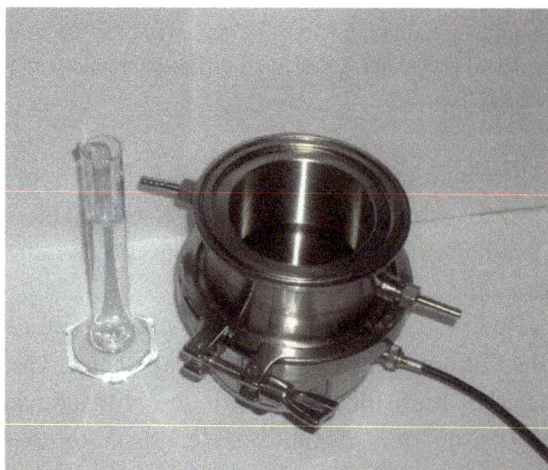

Figure 8: 555-kHz Cup-horn reactor with stainless steel double jacket to cool the solution.

Low and high frequencies can be produced with this equipment, depending on the chosen ceramic and the corresponding imposed frequency.

Interestingly, Naffrechoux *et al.* investigated the sonochemical efficiency of a cup-horn sonoreactor as a function of the liquid height (29–348 mm) and the frequency (22 kHz, 371 kHz, and 504 kHz).[8] The frequency effect was shown to be coupled to liquid height effect. Indeed, acoustic zones, whose limits depend on both transducer diameter and frequency, significantly determined the production of radical species. An increase of ultrasonic frequency resulted in lower acoustic yield and higher sonochemical efficiency. Sonochemical efficiencies obtained at 500 kHz were similar or higher than those at 371 kHz, depending on liquid height. They concluded that the dependence of sonochemical efficiency with liquid height might be firstly attributed to reactor configuration prior to frequency effects.

4.2. Whistle reactors

A whistle reactor is a homogenizer pump that forces the passage of the reaction mixture through a special orifice from which it emerges as a jet which impacts upon a steel blade (Figure 9). This radiation procedure is very advantageous for reactions in liquid/liquid biphasic systems that are efficiently emulsified in the blades. However, this device is not suitable for solid/liquid systems because of the narrow width of the unadjusted whistle to the size of solid particles. Furthermore, the physico-chemical properties of the reagents used must be controlled to maintain the state of tuning fork that can, for example, be quickly eroded by corrosive and cavitation.

4.3. Shape of reactors

The ultrasonic intensity rapidly decreases both radially and axially from the ultrasonic probe. For this reason, the space between the ultrasonic probe and the wall of the reactor must be kept to a minimum, while ensuring that the probe does not touch the container (otherwise the probe might break). Keeping dead zones to a minimum ensures a maximum contact between the sample and the cavitation zones and also among the sample particles, which helps to diminish their size by collisions and hence increases the total area in contact with the solution.

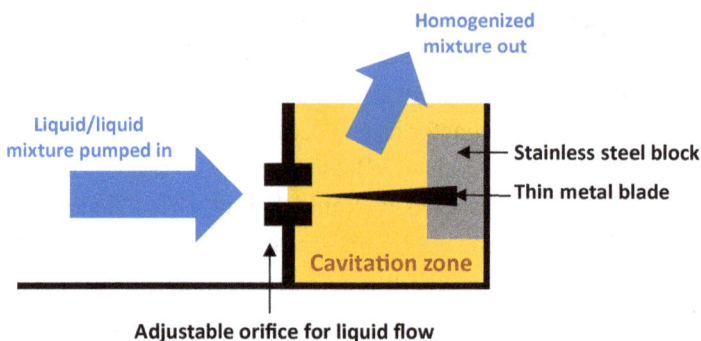

Figure 9: Functional scheme of the liquid whistle.

Figure 10 shows different types of vessels. The influence of their shapes on the extraction of polycyclic aromatic hydrocarbons from sediments under an ultrasonic field has been studied (3- and 6-mm titanium microtip, 22.5 kHz).[9] Sample treatments carried out with reactors D and E led to recoveries below 70% (Figure 10). This low extraction efficiency was linked to the dead cavitation zones due to the distance between the probe and the wall reactor.

Some other shape reactors were designed. For example, a Rosett-type reactor with a flanged lid (Figure 11) allows the propulsion from the end of the probe around the loops of the vessel during the sonication to favor the cooling (if the vessel is immersed in a thermostat bath) and a dynamic and efficient mixing. Duclaux *et al.* reported a more efficient reduction of vermiculite particle size using the Rosett reactor in comparison to a cylindrical reactor due to an additional hydrodynamic cavitation.[10]

Tubular reactors (also called cylindrical tube or sonotube) can effectively transform the longitudinal vibration into the radial vibration and thereby generate ultrasound. Furthermore, ultrasound can be focused to form high-intensity ultrasonic field at the core of the tube. The reactor boasts of simple structure and its whole vessel wall can radiate ultrasound so that the electroacoustic transfer efficiency is high. Z. F. Liang *et al.* reported the design of a tubular reactor with a length up to 2 m.[11] The first advantage is that liquids can pass through it continuously, allowing one to develop a continuous process at a larger scale. The second strength of this

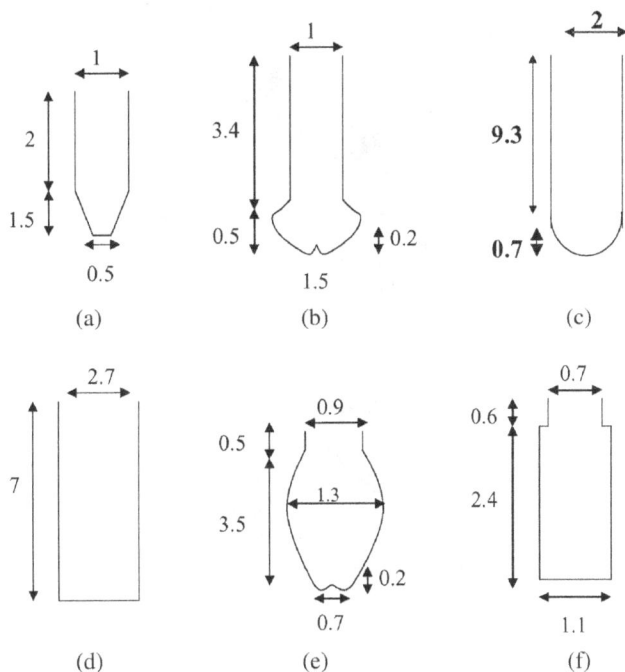

Figure 10: Different vessel designs (measurement in cm). Reprinted with permission from Ref. 9. Copyright (2016) Elsevier.

type of reactor is the limitation of corrosion phenomenon for the equipment, since the maximum of cavitation is observed in the core of the tube and not near to a probe, for example.

Another interesting example is the continuous sonocrystallization in Europe's largest alumina production facility at Aughinish Alumina, Askeaton, Ireland. The process operates continuously at 20 kHz and 10 bar at 70°C using slurry in 6 M NaOH with a throughput of 70 tons per hour (Figure 12).[12]

4.4. Reactor design and scale-up

The design of sonochemical reactors and the rationale for the scale up of successful laboratory ultrasonic experiments are clear goals in sonochemistry.[13,14] Indeed, the progress of sonochemistry in Green Chemistry is

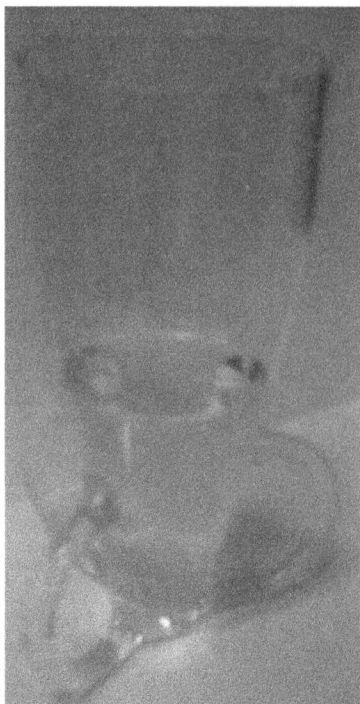

Figure 11:   Rosett-type reactor.

Figure 12:   Large scale continuous crystallization in alumina production facility at Aughinish Alumina, Eire. Reprinted with permission from Ref. 12. Copyright (2016) Elsevier.

dependent upon the possibility of scaling up the excellent laboratory results for industrial use.[15]

"Ultrasonic processing: Now a realistic option for industry" according to T. J. Mason, from Coventry University[16]
"The future contribution of sonochemistry to green and sustainable science is dependent upon the possibility of scaling up excellent laboratory results for industrial use."

In the goal to scale up a sonochemical process, it is essential to first know well the mechanisms involved at the laboratory scale, to determine whether the ultrasonic enhancement is the result of mechanical or chemical effects, as well as the combination of both. The second point to determine is whether the reactor should be of the batch or flow type. Three systems could be envisaged:

- Immerse reactor in a tank of sonicated liquid — indirect irradiation (e.g., flask dipped into an ultrasonic bath).
- Immerse an ultrasonic source directly into the reaction medium — direct irradiation (e.g., probe placed in a reaction vessel).
- Use reactor constructed with vibrating walls — direct irradiation (e.g., a tube operating through radial vibrations).

The use of ultrasound on an industrial scale remains currently reserved for welding (Sonics&Materials Inc., Telsonic Ultrasonics, Eurosonic Ultrasonic Technology, etc.) and cleaning of industrial components (Guyson International, Société Nouvelle N.P.I 25, Unitech Annemasse, Omegasonics, etc.).

Since the example of a batch industrial ultrasonic reactor designed and built in Romania for the extraction and preparation of tinctures from various herbs (Figure 13),[17] decreasing from 28 days to 10 hours of the extraction process, some recent example streams have been developed:

⇨ **Ultrasonic extraction:** The method developed by the company *Hielscher Ultrasonics* is used to extract organic compounds contained

Figure 13: Ultrasonic 1,000 l batch extractor. Reprinted with permission from Ref. 17. Copyright (2016) Elsevier.

in plants and seeds.[18] The process of the company Industrial Sonomechanics showed that ultrasound promotes the extraction of oils through the walls of the cells and allows a sufficient mass transfer between the oil and the alcohol to accelerate the transesterification reaction.[19] The ultrasonically assisted extraction (UAE) was applied for phytopharmaceutical extraction industry and represents a versatile technique that can be used both on a small and large scale.[17,20]

⇨ **Simultaneous generation and deposition of nanoparticles on fabrics:** A pilot installation was designed for the scale up of sonochemically assisted coating of textile fabrics with various kinds of nanoparticles. This continuous process has produced biocidal cotton bandages containing 0.65 wt.% and 1.50 wt.% of CuO, showing a good antimicrobial properties against *E. coli* bacteria.[21,22] The nanoparticles are embedded into the fibers by the cavitation jets, which are formed by asymmetrically collapsing bubbles in the presence of a solid surface and are directed towards the surface of textile at very high velocities. Fabrics coated with ZnO nanoparticles were also produced in continuous flow.[22]

⇨ **Ultrasonic crystallization:** The *Termed Solution Atomization and Crystallization by Sonication* application (SAX™) enabled the large scale production of corticosteroids, including Budesonide a synthetic anti-inflammatory drug administered by inhalation in the treatment of asthma and hay fever (Figure 14).[23] The most industrially developed process from SAX method is named UMAX® for *Ultrasound Mediated Amorphous to Crystalline transition*, incorporating spray drying and

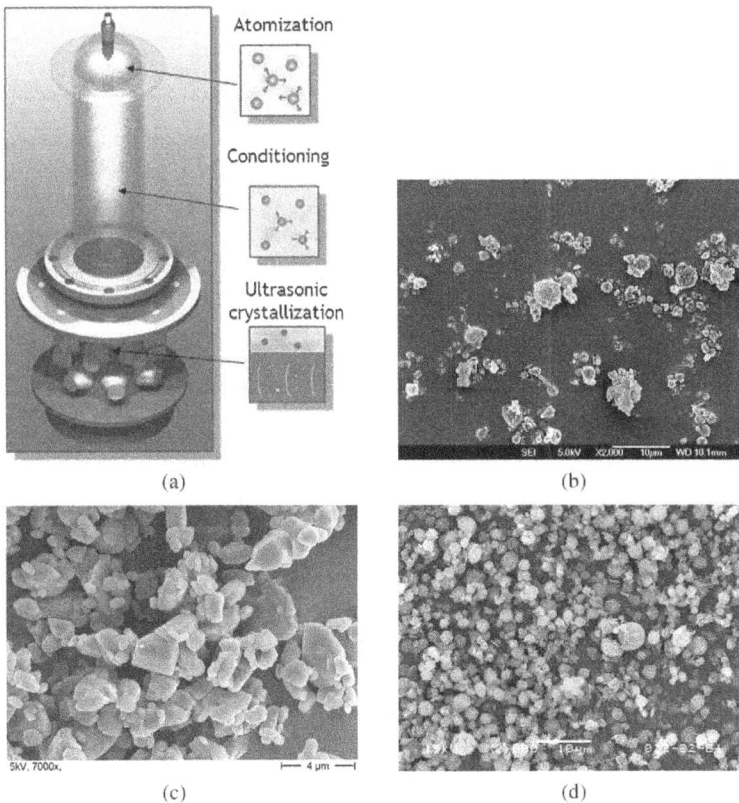

(a) (b) (c) (d)

Figure 14: Ultrasonic crystallization: (a) schematic SAX process; (b) cortecosteroid prepared normally; (c) cortecosteroid prepared normally then micronized; (d) cortecosteroid prepared by the UMAX® system. Reprinted with permission from Ref. 16. Copyright (2016) Elsevier.

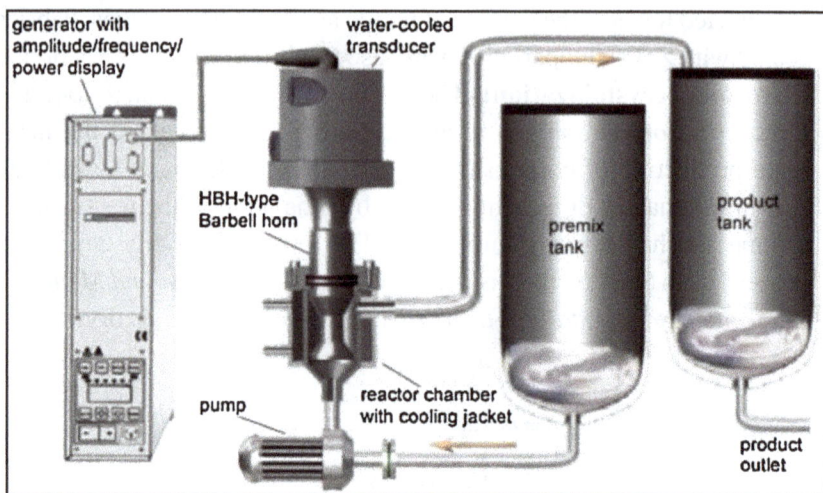

Figure 15: Operating of the BHUT system: The generator excites vibration in the water-cooled piezoelectric transducer; the Barbell horn amplifies the vibration amplitude and delivers the ultrasonic energy to the liquid premix, which is continuously pumped through the reactor chamber from the premix tank into the product tank. Reprinted with permission from Ref. 24. Copyright (2016) Elsevier.

ultrasonication now used to manufacture a wide range of drugs materials for asthma and chronic obstructive pulmonary disease.[16]

⇨ **Continuous-flow production of pharmaceutical nanoemulsion under ultrasound:** The *Barbell Horn Ultrasonic Technology* (*BHUT*) permits constructing bench and industrial-scale processors capable of operating at high ultrasonic amplitudes. For example, it will be used to produce a high-quality MF59®-analog pharmaceutical nanoemulsion at a large scale (Figure 15).[24]

Some other examples reporting the use of sonochemistry at an industrial scale may be mentioned here (Figure 16). Nevertheless, the application of ultrasound to the industrial scale, in particular for chemical reactions, is still limited by some technical difficulties and implementation. The main technological challenges are about the improvements in terms of (i) energy distribution within the reactor, and (ii) reproducibility of experiments.

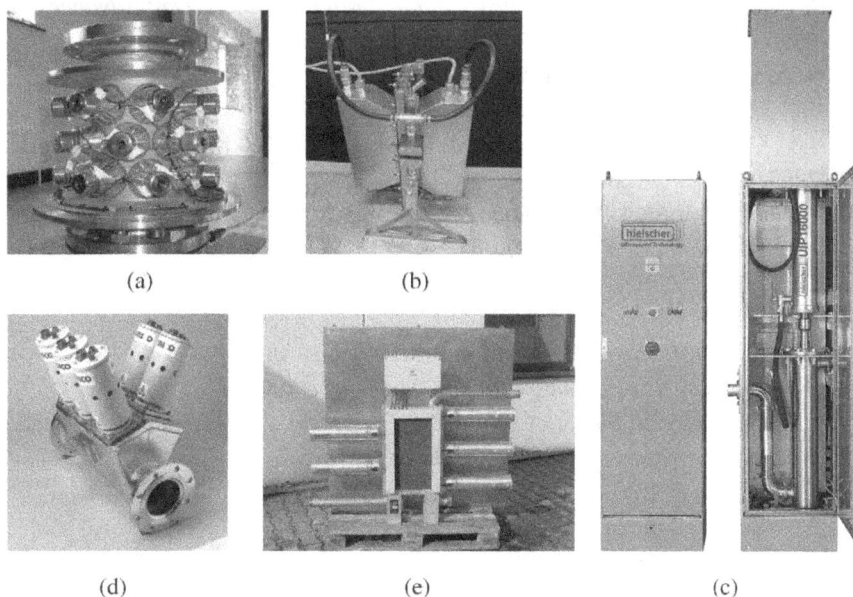

Figure 16: Ultrasonic flow treatment modules: (a) prosonitron from Prosonix, UK; (b) Nearfield Acoustic Processor (NAP reactor) from Advanced Sonic Processing Systems, USA; (c) UIP16000 from Hielscher, Germany; (d) V5 reactor from Enpure (UK); (e) sludge disruptor from Ultrawaves, Germany. Reprinted with permission from Ref. 16. Copyright (2016) Elsevier.

Nowadays, academics, equipment designers and manufacturers are actively engaged in developing larger scale sonochemical processing that represent a crucial parameter in the final exploitation of green sonochemistry.[12]

References

1. L. H. Thompson, L. K. Doraiswamy, *Ind. Eng. Chem. Res.* **1999**, *38*, 1215–1249.
2. R. L. Hunicke, *Ultrasonics* **1990**, *28*, 291–294.
3. B. Pugin, *Ultrasonics* **1987**, *25*, 49–55.
4. T. J. Mason, *Sonochemistry*, Chemistry Primers, Oxford, 1999, 92.
5. V. S. Moholkar, S. P. Sable, A. B. Pandit, *AIChE J.* **2000**, *46*, 684–694.
6. Z. Wei, J. A. Kosterman, R. Xiao, G.-Y. Pee, M. Cai, L. K. Weavers, *Ultrason. Sonochem.* **2015**, *27*, 325–333.

7. J.-P. Bazureau, M. Draye, *Ultrasound and Microwaves: Recent Advances in Organic Chemistry*, Transworld Research Network, Kerala, 2011, 241.

8. S. de La Rochebrochard d'Auzay, J.-F. Blais, *Ultrason. Sonochem.* **2010**, *17*, 547–554.

9. J. L. Capelo, M. M. Galesio, G. M. Felisberto, C. Vaz, J. C. Pessoa, *Talanta* **2005**, *66*, 1272–1280.

10. F. Ali, L. Reinert, J.-M. Levêque, L. Duclaux, F. Muller, S. Saeed, S. S. Shah, *Ultrason. Sonochem.* **2014**, *21*, 1002–1009.

11. G. P. Zhou, Z. F. Liang, Z. Z. Li, Y. H. Zhang, *Chin. Sci. Bull.* **2007**, *52*, 1902–1905.

12. T. J. Mason, *Ultrason. Sonochem.* **2007**, *14*, 476–483.

13. P. R. Gogate, A. B. Pandit, *Ultrason. Sonochem.* **2004**, *11*, 105–117.

14. P. R. Gogate, V. S. Sutkar, A. B. Pandit, *Chem. Eng. J.* **2011**, *166*, 1066–1082.

15. P. Cintas, *Ultrason. Sonochem.* **2016**, *28*, 257–258.

16. C. Leonelli, T. J. Mason, *Chem. Eng. Process.* **2010**, *49*, 885–900.

17. M. Vinatoru, *Ultrason. Sonochem.* **2001**, *8*, 303–313.

18. Hielscher Ultrasonics, Webpage [online, available on March 2016]: http://www.hielscher.com/ultrasonics/extraction_01.htm.

19. Industrial Sonomechanics, Webpage [online, available on March 2016]: http://sonomechanics.com/applications.

20. M. Vinatoru, *Ultrason. Sonochem.* **2015**, *25*, 94–95.

21. A. V. Abramova, V. O. Abramov, A. Gedanken, I. Perelshtein, V. M. Bayazitov, *Beilstein J. Nanotechnol.* **2014**, *5*, 532–536.

22. V. O. Abramov, A. Gedanken, Y. Koltypin, N. Perkas, I. Perlshtein, E. Joyce, T. J. Mason, *Surf. Coat. Tech.* **2009**, *204*, 718–722.

23. G. Ruecroft, *Eur. Ind. Pharm.* **2009**, *16*, 16–17.

24. A. S. Peshkovsky, S. Bystryak, *Chem. Eng. Process.* **2014**, *82*, 132–136.

Chapter 5

Applications in Green Chemistry

This chapter presents some eco-friendly applications in different fields of chemistry such as catalysis and organic chemistry, materials preparation, polymer chemistry, biomass conversion, extraction and environmental remediation. More than a complete review of the literature, we selected specific examples to highlight how sonochemistry can efficiently contribute to green chemistry. A section of this chapter is particularly focused on innovative combinations of ultrasound with other technologies such as microwave, electrochemistry, uses of ionic liquids or enzymes. The final part will be dedicated to some comments about scale-up and industrial applications.

1. Sonocatalysis and Organic Sonochemistry

High-power ultrasound can generate cavitation within a solvent and through cavitation provide a source of energy which can be used to enhance a wide range of organic reactions. Such uses of ultrasound are grouped under the general name "organic sonochemistry" and/or "sonocatalysis" when a catalyst is involved. Several reviews have covered the entire scale of applications of ultrasound in the area.[1–5] In this section, we have selected some interesting examples focused on applications in organic synthesis where ultrasound seems to provide a distinct alternative to other and more traditional techniques of improving reaction rates and product yields. In some cases, it can also provide new synthetic pathways.

The first examples involving ultrasound in organic sonochemistry were not very relevant since they were performed in homogeneous solutions, limiting their interests in these conditions. Then, some researchers use low frequency ultrasound in biphasic systems as a mixing means for the heterogeneous chemistry. However, the first important breakthrough to modern sonochemistry is attributed to the Jean-Louis Luche's group, when they reported in 1980 the rapid sonochemical preparation of various types of organolithium compounds and their uses as reactants for a Barbier coupling with carbonyl compounds (Scheme 1).[6] Here, the use of an ultrasonic bath (50 kHz, 60 W) led to an important improvement of the reactions yield, probably due to the mechanical effects of ultrasound that altered the metal surface of the catalyst.

In 1984, another great discovery was reported by the Takashi Ando's group with the first example of **sonochemical switching**.[7] When benzyl bromide was treated with potassium cyanide and alumina in toluene under mechanical stirring at 50°C, only a mixture of *o*- and *p*-benzyltoluene was obtained in 83% yield through a Friedel–Crafts reaction (Scheme 2). However, when the same reaction mixture was irradiated by ultrasound (ultrasonic bath, 45 kHz, 200 W) at 50°C, the substitution took place to afford benzyl cyanide in 76% yield. Here, the ultrasonic irradiation completely switched the reaction pathway from the Friedel–Crafts reaction to nucleophilic substitution.

This change of reactivity was first assigned to a poisoning of the alumina by potassium cyanide which may increase under sonication. However, benzyl cyanide may be synthesized through a Single Electron Transfer (SET) reaction as sonication should increase the yield of free

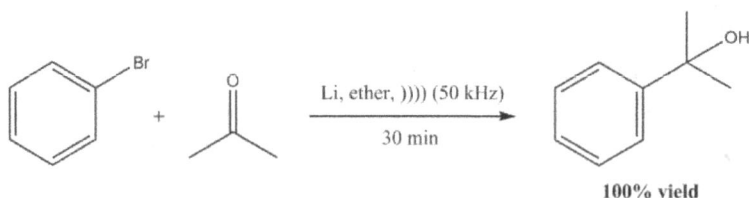

Scheme 1: Barbier coupling performed under ultrasound.[6]

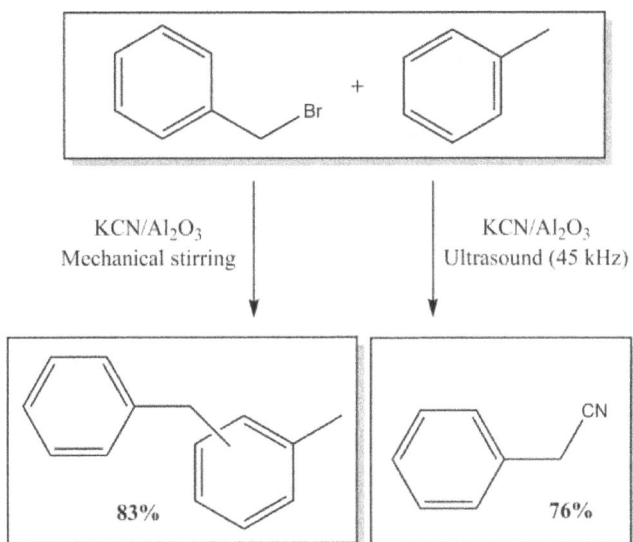

Scheme 2: Sonochemical switching reported by the Ando's group.[7]

radical species compared to classical stirring. Even if the explanation remains a little bit controversial, this work is considered as the first example of **true sonochemistry**.

"True sonochemistry" versus "False sonochemistry"[3,8,9]

Jean-Louis Luche made the important observation that sonochemistry applications could be subdivided into reactions which were the result of "true" and "false" effects. These terms referred to real chemical effects induced by cavitation and the effects that could be mainly ascribed to the mechanical impact of bubble collapse.

The **false sonochemistry** regroups the reactions that take benefits from the physical effects of ultrasound such as reduction of particles size, microemulsion and mass transfer phenomena. "False sonochemistry" is not a pejorative term at all and in many cases, it can lead to important improvements in terms of reaction yield, kinetics or even feasibility, especially for heterogeneous chemistry.

The **true sonochemistry** is related to the chemical effects of ultrasound and often, to radical chemistry.

More precisely, chemical reactions have been separated into three main categories[3,10]:

⇨ **Type I:** homogeneous reactions taking place through radical or radical-ion intermediates through a SET increased under ultrasound. The excited species may also include coordinatively unsaturated metal compounds.

⇨ **Type II:** heterogeneous reactions involving ionic reactions. Mechanical effects (surface cleaning, particle size reduction, enhanced mass transfer, etc.) affect directly the kinetics and the yields of organic reactions. Related to false sonochemistry, these effects could be attained with efficient methods of stirring, milling, friction or grinding.

⇨ **Type III:** heterogeneous systems capable of following either ionic or radical mechanism, the latter favored under ultrasound (SET pathway). This kind of reaction is affected by both the chemical and physical effects of ultrasound. They are sometimes called ambivalent reactions. The sonochemical switching proposed by Ando *et al.* is an example of type III reaction.[7]

Experimentally, it is often difficult to clearly identify the real effects brought by ultrasound irradiation and to exactly understand (and prove) how it can affect the reactivity. What is important is the systematic comparison with silent conditions (without ultrasound, that is to say the "blank reaction"). In terms of green chemistry, an improvement of yield, selectivity or rate under ultrasound remains beneficial, even if the mechanisms are not always elucidated[11]: (i) the increase of the yield automatically leads to a decrease of the E-factor (amount of waste); (ii) the increase of the selectivity generally limits the importance of the costly workup (use of solvent for extraction, separation, recrystallization, etc.); (iii) the improved kinetics are favorable to decrease the energy consumption through the reduction of reaction times.

We report in the following pages some recent examples that we selected to illustrate the use of ultrasound as a green technology in organic chemistry and catalysis areas. More complete review of the literature has been made recently in different interesting articles and chapters/books.[1–5,12]

> **A list of interesting effects under ultrasound...**
>
> Here is the list of the main effects reported in the literature when ultrasound irradiation is beneficial for a chemical reaction:
>
> ⇨ Decrease in reaction times;
> ⇨ Increased reaction yields;
> ⇨ Use of less harsh conditions (lower temperature, etc.);
> ⇨ Change of the reaction pathway, new reactivity;
> ⇨ Reduced amounts of catalyst in the reaction or no required catalyst;
> ⇨ Forced degassing for reactions involving gaseous products;
> ⇨ Use of raw reagent or at a technical grade;
> ⇨ Activation of metals and solids;
> ⇨ Improved reactivity of the reactants/catalysts;
> ⇨ Generation of useful reactive species.
>
> All these beneficial effects provided by ultrasound are in accordance with the twelve principles of green chemistry (see Chapter 1, page 7).

1.1. Examples of the synthesis of fused heterocycles

Some impressive results were reported under ultrasound for the synthesis of fused heterocycles, especially for medicinal chemistry and drug discovery processes.[13] The main gain is the reduction of reaction times from days or hours to minutes. In many cases, sonochemical synthesis also provides higher yields, lower cost, easy workups and greater purity as compared to conventional thermal methods.

For example, Nikpassand *et al.* reported the synthesis of a novel fused tetracyclic derivatives of pyrazolopyridines by reacting 5-amino-3-methyl-*1H*-pyrazole, *2H*-indene-1,3-dione and aldehyde using an ultrasonic bath equipped with two transducers at 45 kHz, with an input power of 305 W (Scheme 3).[14] In this case, even if the specific role of ultrasound was not investigated, the ultrasonic route led to the reduction of reaction time from 4–5 hours to 4–5 min with an important improvement of yields.

Another recent example was reported for the one-pot synthesis of spiro[indoline-3,4′-pyrano[2,3-c]pyrazole] derivative catalyzed by

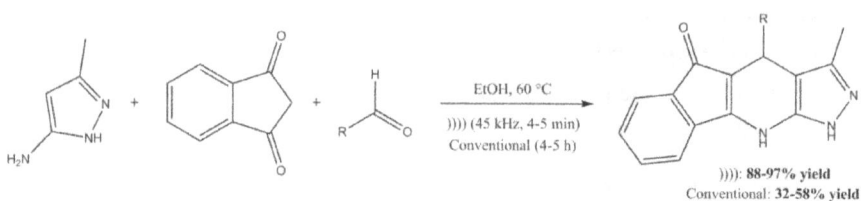

Scheme 3: Synthesis of fused tetracyclic derivatives of pyrazolopyrines under ultrasound (with R = o-NO$_2$C$_6$H$_4$, m-NO$_2$C$_6$H$_4$, p-NO$_2$C$_6$H$_4$, m-BrC$_6$H$_4$, m-ClC$_6$H$_4$, p-ClC$_6$H$_4$, 2,4-Cl$_2$C$_6$H$_3$, 3-indoyl, 5-NO$_2$-2-furyl, etc.).[14]

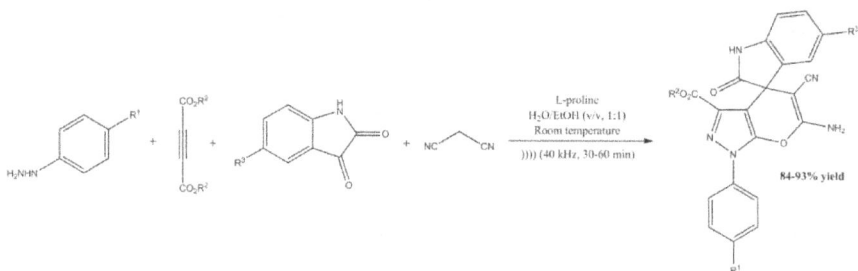

Scheme 4: L-proline-catalyzed one-pot four component synthesis of spiro[indoline-3,4'-pyrano[2,3-c]pyrazole] derivatives in water/ethanol under ultrasound.[15]

L-proline under 40 kHz (nominal power of 250 W) in water/ethanol mixture as solvent (v/v, 1:1) at room temperature (Scheme 4).[15] Reaction yields between 84% and 91% were obtained in these conditions in only 60 min. However, the reaction also occurred in silent conditions but led to lower yields (81%) in a longer reaction time (480 min).

Cella and Stefani also reported the examples from the literature of heterocycles chemistry.[16] In the major cases, the effects of ultrasound are not explained and most of the papers attributed the positive improvements brought by ultrasound to the "cavitation phenomenon" without more explanations. The efficient micromixing and mass transfer implied by sonication seem to be the main reasons of these excellent yields. Often, no solvent is required under ultrasound, leading to interesting syntheses in term of green chemistry.

For example, Li *et al.* reported the synthesis of oxabicyclic alkenes by ultrasound-promoted Diels–Alder cycloaddition of furano dienes (Scheme 5).[17] When furans and dimethyl acetylenedicarboxylate

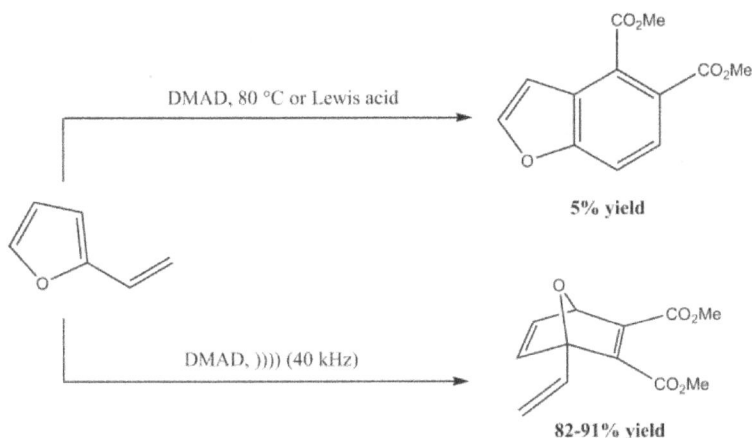

Scheme 5: Diels–Alder cycloaddition from furans and DMAD.[17]

(DMAD) were irradiated under ultrasound (40 kHz, 160 W, ultrasonic bath) without solvent, the desired cycloadducts were obtained with 82–91% yields. Interestingly, under silent conditions, even at 80°C, the only product formed in a low yield was the exo-cyclic diene. In this case, the reactivity and probably the mechanism pathway were modified under ultrasound.

Numerous examples of Diels–Alder cycloadditions performed under ultrasound were reported in the literature.[18–21] Already 20 years ago, some researchers were talking about the *"intriguing problem in organic sono-chemistry of the Diels–Alder reaction"*, explaining that the thermal effects of cavitation did not seem to be directly involved in the mechanism.[22] They proposed a redox process between the diene and dienophile, giving radical-ion intermediates.

Safari *et al.* reported the synthesis of 2-amino-4,8-dihydropyrano [3,2-*b*]pyran-3-carbonitrile derivatives *via* three-component reaction of arylaldehydes, malononitrile and kojic acid in water at 50°C under ultrasound irradiation (40 kHz, nominal power of 200 W, ultrasonic bath).[23] The reported method provided several advantages in terms of green chemistry point of view with shorter reaction time, excellent yields (70–88%), no requirement base, earth metal Lewis acid and simple workup procedure (Scheme 6).

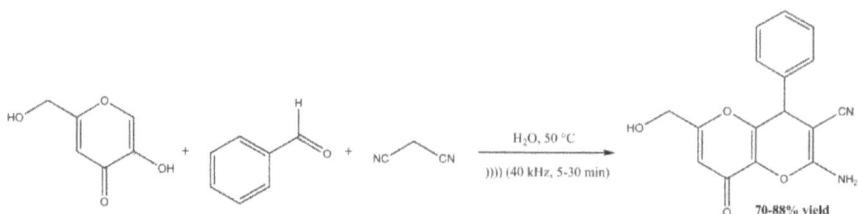

Scheme 6: Synthesis of 2-amino-4,8-dihydropyranol[3,2,*b*]pyran-3-carbonitriles in water under ultrasound.[23]

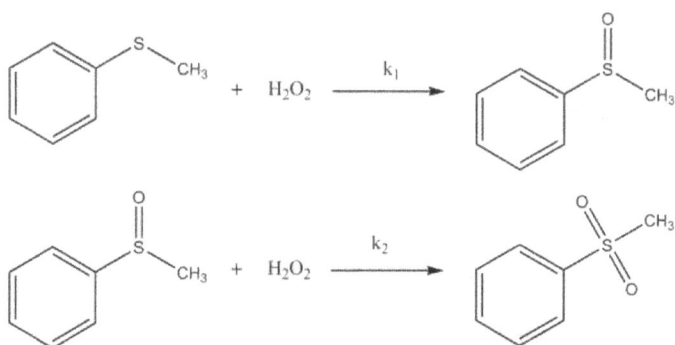

Scheme 7: Oxidation of methyl phenyl sulfide to methyl phenyl sulfoxide and methyl phenyl sulfone using H_2O_2 under ultrasound at room temperature for 2 hours.[30]

1.2. Examples of oxidation reactions

Ultrasound irradiation was also widely used for selective oxidation reactions in the presence of nanoparticles,[24–26] metal oxides[27] or other catalysts.[28,29] For example, Pandit *et al.* reported the selective oxidation of sulfides to sulfoxides under low frequency ultrasound (22 kHz, output power of 120 W, ultrasonic bath).[30] Hydrogen peroxide, a clean oxidant producing only water as a by-product, was used in aqueous medium instead of the classical method performed in methanol (Scheme 7). The easy separation of products and the recyclability of reagents showed some interesting results in terms of green chemistry. A detailed study of the effect of various parameters has been done. The presence of ultrasound significantly enhanced the rate as well as selectivity of the oxidation reaction, even at room temperature, as compared to the conventional techniques. A molar ratio of 1:1 of methyl phenyl sulfide to H_2O_2 is advantageous for a maximum selectivity of sulfoxide (100%) at lower conversions of sulfide (93%). Interestingly, authors demonstrated that the optimum conditions found in laboratory-scale experimentation were equally applicable in a scaled up version of the system.[30]

Queneau *et al.* showed the selective oxidation of sucrose (primary hydroxyl group) in homogeneous aqueous medium by NaOCl/TEMPO ((2,2,6,6-tetramethylpiperidin-1-yl)oxyl) under ultrasound (20 kHz, electric power of 300 W).[31] Here, authors highlighted that the pretreatment of the oxidants by ultrasound prior addition of the carbohydrate substrate led to a significant rate increase compared to the classical conditions.

In the same objective, we recently proposed an efficient, eco-friendly and selective method to oxidize D-glucose with aqueous hydrogen peroxide in the presence of iron(II) sulfate as a cheap catalyst (Scheme 8).[32] Indeed, the production of the biodegradable gluconic acid (about 100,000 tons each year around the world)[33] is industrially used as a water-soluble chelating and acidifying agent for cleaning applications, and as an additive in food and pharmaceutical industries. In the medical field, gluconic acid is also used as dietary supplement to prevent cancer.[34]

Using low frequency ultrasound (20 kHz, $0.25 \text{ W} \cdot \text{mL}^{-1}$) at 25°C as activation method during short times (15 min), we selectively converted D-glucose into gluconic acid with excellent yields (97% with temperature of only 22.5°C).[32] To better understand why the sonochemical oxidation of D-glucose was more efficient than traditional heating, we studied the mechanism of the reaction. In the presence of H_2O_2 and ferrous ions at acidic pH, it is known that we can directly produce hydroxyl radicals *via* a Fenton process (production of $HO^\bullet$ species). First, in the presence of *tert*-butyl alcohol as radical scavenger in the optimized conditions, the glucose conversion and yield in gluconic acid was less than 1%, showing

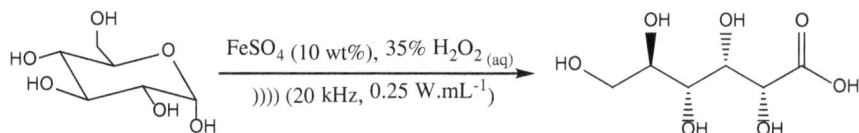

Scheme 8: Oxidation of D-glucose into gluconic acid under ultrasonic conditions.[32]

that the presence of HO• radicals was essential in this reaction. In addition, based on the measurement of radical production (KI dosimetry method), we clearly demonstrated the role of radical in the oxidation of D-glucose into gluconic acid, improved under ultrasound *via* a sono-Fenton process.[32] Interestingly, we also demonstrated in this study that low frequency ultrasound activation led to an energy consumption 4.5 times less compared to conventional heating for the similar yields, representing a serious advantages in terms of green chemistry.

1.3. Examples of hydrogenation reactions

Several examples were also reported under ultrasound for hydrogenation reactions in the presence of heterogeneous catalysts. Disselkamp *et al.* compared conventional heating and ultrasound activation (20 kHz, 90/190 W, ultrasonic probe) in the hydrogenation of 3-buten-1-ol aqueous solution under 6.8 atm of hydrogen using Pd-black powder.[35] Authors determined that ultrasound creates catalyst sites enhancing the *cis*-to-*trans* 2-buten-1-ol ratio from 0.25 to 0.55. In addition, comparing the total isomerization to hydrogenation ratio (*cis*- plus *trans*-2-buten-1-ol to 1-butanol ratio) for ultrasound-assisted and conventional catalysis reveals a ~five-fold enhancement in isomerization relative to the more energetically favored hydrogenation due to the application of ultrasound. In this study, they showed the conversion of a terminal alkene into an internal one forming both *cis* and *trans* isomers with an enrichment ~five-fold under ultrasound compared to conventional catalysis.

Tripathi *et al.* reported the selective hydrogenation of C-5 acetylene alcohols such as 2-methyl-3-butyn-2-ol using Lindlar catalysts under 40 kHz, 380 kHz and 850 kHz ultrasonic irradiations (Scheme 9).[36] Low frequency (40 kHz) led to the best results through the better dispersion of the catalyst provided by the mechanical effects of ultrasound. With high frequencies, a supplementary mechanical stirring is necessary to improve the mass transfer during the reaction.

Interestingly, the Cravotto's group at the University of Turin in collaboration with Danacamerini SAS developed an own-built reactor for hydrogenations under moderate pressure, up to 7 bars (Figure 1).[37] Indeed, sonochemical hydrogenations can be generally easily carried out at room

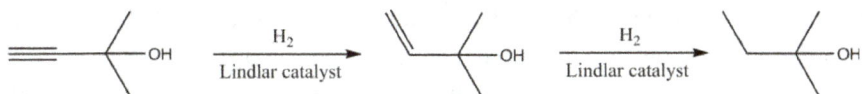

Scheme 9: Hydrogenation of 2-methyl-3-butyn-2-ol using the Lindlar catalyst using conventional stirring and three different ultrasonic frequencies (Experimental conditions: 1 bar H_2, 2.5 hours, 20°C, 100 mg Lindlar catalyst for 20 mmol substrate in 20 mL MeOH).[36]

Figure 1: Ultrasound-reactor for low-pressure hydrogenations. Reprinted with permission from Ref. 37. Copyright (2016) Elsevier.

temperature, thanks to the acoustic cavitation that maximizes the mass transfer and the activation of the catalyst, especially at low frequencies.[38]

1.4. Other examples of heterogeneous catalysis

Khorshidi reported the oxidation by $KBrO_3$ of naphthalene to phthalalde-hyde under ultrasound irradiation (ultrasonic bath, 24 kHz/50–60 kHz, power of 250 W) in the presence of ruthenium nanoparticles supported on mesoporous MCM-41 as catalyst.[39] In this case, the ultrasound irradiation allowed the acceleration of the oxidation reaction to afford the desired products in good yields (69%). The recovered catalyst retained activity for successive runs, but a continuous change in the nature of its active sites was observed. The efficiency of the reaction was tested at 24 kHz and 50–60 kHz, leading to the same yield in aldehyde for a similar power density.

Safari and Javadian prepared magnetic Fe_3O_4–chitosan nanoparticles for the one-pot synthesis of 2-amino-*4H*-chromenes by condensation of aldehydes with malononitrile and resorcinol under ultrasound irradiation (ultrasonic bath, 35 kHz, nominal power of 200 W).[40] Here, ultrasound was used both for the preparation of catalyst (better dispersion) and the reaction (best yields and lower reaction times, Table 1). Indeed, under ultrasound, the yield was improved (quantitative reaction) and the reaction time reduced to only 20 min (Table 1, entry 5). Interestingly, as a green chemistry point of view, the magnetic catalyst was easily recovered and recycled for six successful runs.

Nasrollahzadeh *et al.* reported another type of reusable catalyst, Natrolite nanozaolite, for the *N*-sulfonylation of primary and secondary amines, using ultrasound (25 kHz and 40 kHz, output power of 250 W) at room temperature.[41] In the optimized conditions, the reaction was performed in ethanol, at 40 kHz, in only 4 min with 96% yield. The heterogeneous catalyst was also recycled and efficiently reused.

Another recent example was reported by Mirza-Aghayan *et al.* using graphite oxide as catalyst for the ultrasound-assisted oxidative amidation

Table 1: Synthesis of 2-amino-4*H*-chromenes under silent and ultrasonic conditions (Experimental conditions: 1 mmol aldehyde, 1 mmol malononitrile, 1 mmol sesorcinol, Fe_3O_4–chitosan catalyst (0–0.20 g) in water under reflux (silent conditions) or at 50°C (ultrasonic conditions)).[40]

Entry	Catalyst (g)	Yield without ultrasound (%)	Reaction time without ultrasound (min)	Yield with ultrasound (%)	Reaction time with ultrasound (min)
1	0	71	65	82	30
2	0.05	78	47	87	25
3	0.10	85	49	88	23
4	0.15	90	31	99	20
5	0.20	90	31	99	20

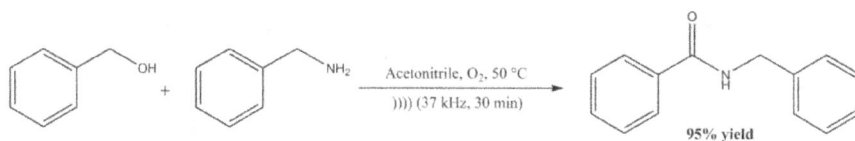

Scheme 10: Oxidative amidation of benzyl alcohol with benzyl amine.[28]

of benzyl alcohols.[28] For example, the synthesis of *N*-benzylbenzamide was performed in acetonitrile using an ultrasonic bath (37 kHz) at 50°C (Scheme 10). The yield was 55% after 90 min when the reaction was performed under argon atmosphere, but it reached 95% after 30 min when the reaction was saturated with O_2 gas, showing the important role of molecular oxygen in the reaction. Authors admit that the mechanism under ultrasound was not very clear, but they showed that the reaction is not efficient under silent conditions (18% yield after 48 hours). In addition, they recycled and reused the catalyst for six runs, without a loss of activity.

Cravotto and Cintas discussed the improvements that ultrasound can provide on the physical and chemical transformations by means of efficient stirring, dissolution, mass and heat transfers and reagents sonolysis, which all arise from the cavitational collapse.[10] Indeed, the majority of the examples reported in sonocatalysis field can be attributed to **false sonochemistry**, related to mechanochemistry in modern synthesis, illustrating the advantages of using pressure waves in chemistry. Thus, most of results in organic sonochemistry can be explained by the effect of mechanochemical energy or micromixing and emulsification effects, both provided by cavitation. The examples involving **true sonochemistry** and a single electron transfer (sonochemical switching) are much less common in the literature for the moment.

2. Sonochemical Preparation of Materials

The radical and mechanical effects of ultrasound (sonochemistry, related to cavitational chemistry)[42] have been widely investigated for the preparation of catalysts or specific materials. The subject was reviewed, highlighting the fields where ultrasound was expanded: modification of solids,

Materials Applications of Ultrasound

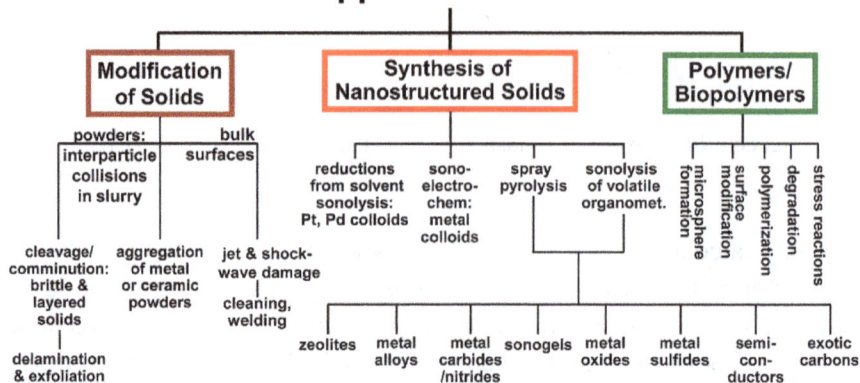

Figure 2: Materials applications of ultrasound. Reprinted with permission from Ref. 44. Copyright (2016) Royal Society of Chemistry.

synthesis of nanostructured solids and polymers/biopolymers (Figure 2).[43,44] Two specific sections are dedicated in this chapter to polymers and biopolymers. Here, we provide some recent examples to show how sono-chemistry can provide eco-friendly solutions for modification of solids or the synthesis of nanostructured solids.

2.1. Modification of solids

Historically, ultrasonic cavitation at metal surfaces was of uppermost importance in metal activation, especially for the formation of organome-tallic reagents and metal-assisted couplings.[45] The main effects provided by ultrasound are the decrease in particle size and the creation of fresh metal surfaces.[46]

Metal and composite nanomaterials were functionalized under low frequency ultrasound. Cubic faced centered (Al, Ag, Cu) or hexagonal (Zn) structured metal plates were strongly attacked. A macroscopic obser-vation of the surface showed strong but homogeneous modifications. Scanning Electron Microscopy (SEM) revealed the development of a surface roughness, appearing after only 1 min of ultrasonic treatment and increasing with sonication time. The treated surface of copper plates was cleaned from its natural oxide layer (Figure 3). The treated surface of silver plate turned white and tarnished after 5 min sonication.[47,48]

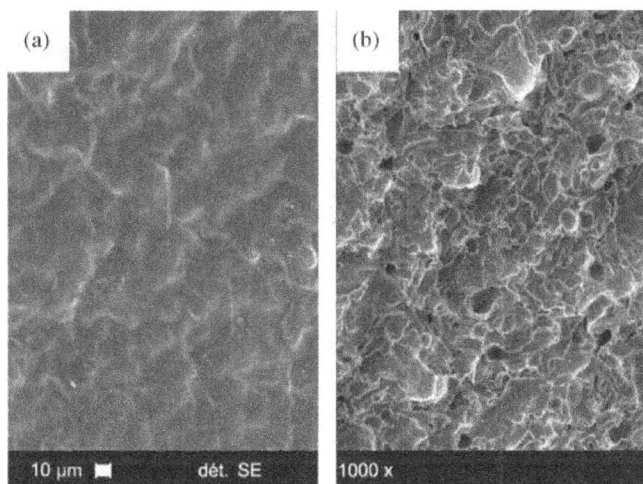

Figure 3: SEM images of the surface of a Cu plate after 5 min (a) and 10 min (b) of ultrasonic treatment (20 kHz, 40 W·cm^{-2}). Reprinted with permission from Ref. 48. Copyright (2016) Elsevier.

Amanov *et al.* reported the nanocrystalline surface modification of magnesium alloy under 20 kHz ultrasonic irradiation.[49] They observed some changes in the tribological properties after this treatment: lower friction coefficient, gradual increase of the surface hardness, etc.

Mizukoshi *et al.* showed the successful immobilization of Pt, Au and Pd nanoparticles onto TiO_2 under ultrasound (200 kHz, input power of 200 W, 6 W·cm^{-2}).[50] Sonication times were comprised between 5 min and 60 min under an air atmosphere. Transmission electron microscopy (TEM) showed that smaller Pt nanoparticles (2.1 nm as average diameter) were immobilized on the surface of larger TiO_2 particles without any aggregation after ultrasonic treatment (Figure 4). The average diameters of the supported Pt nanoparticles were almost the same in both sono-chemical and impregnation methods, but the Pt size distribution of the conventional one was slightly wider. The number of supported Pt nano-particles by the conventional method was much smaller than by the sono-chemical method. In addition, the surface of TiO_2 looks smoother than that of the sonochemical products. The roughness of the sonicated surface was due to the effects of the ultrasound. Interestingly, the amount of

73

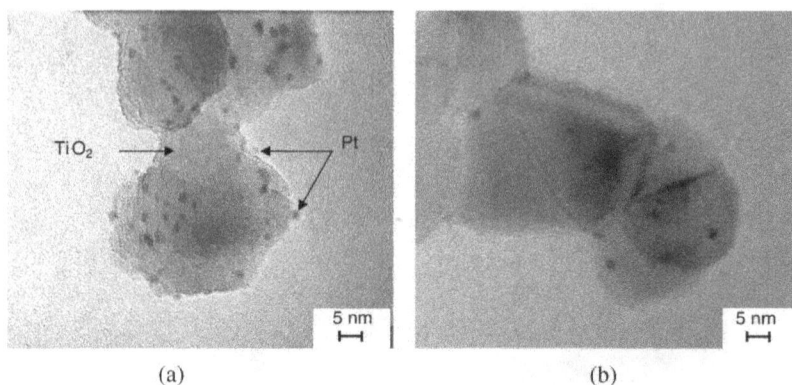

Figure 4: TEM images of Pt/TiO$_2$ catalysts prepared by sonochemical method (a) and by the impregnation method (b). Reprinted with permission from Ref. 50. Copyright (2016) Elsevier.

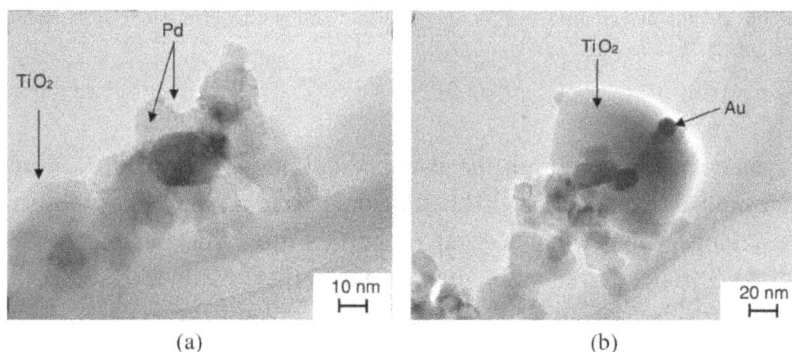

Figure 5: TEM images of sonochemically prepared Pd/TiO$_2$ and Au/TiO$_2$. Reprinted with permission from Ref. 50. Copyright (2016) Elsevier.

loaded Pt was dependent upon the sonication time, allowing the control of Pd loading as a function of reaction time.

Pd/TiO$_2$ and Au/TiO$_2$ were also successfully prepared by similar protocols (Figure 5).

We recently reviewed the use of ultrasound for treating clays.[51] In this field, sonochemistry is essentially used to improve particles reduction and dispersion, to modify the textural properties of clays, to help the intercalation and adsorption phenomenon and to favor the catalytic activity in selected reactions.

For example, 20 kHz ultrasound treatment (350 W output, probe) was applied to improve the exfoliation of vermiculite, one of the most used mineral materials in the R&D solutions for its applications as adsorbent for water treatment, design of nanocomposite materials, making of gaskets for sealing technology, lightweight porous filler in the production of heat-insulating refractory components, etc.[52] Indeed, micron and submicron-size platelets of vermiculite were prepared in water or in aqueous solution of H_2O_2, with the particle size distributions dependent on the treatment time. The modification of the surface charge of the particles brought out by zeta potential's decrease together with sonication time was enhanced by soni-cation treatments in hydrogen peroxide instead of water. In this case, the sonication time was relatively long (until 12 hours, Figure 6).

Belova *et al.* reported the intercalation of gold nanoparticles into multilayered Na^+-montmorillonites under ultrasound (20 kHz, 500 W).[53] All analyses — Brunauer–Emmett–Teller (BET) surface area analysis, TEM,

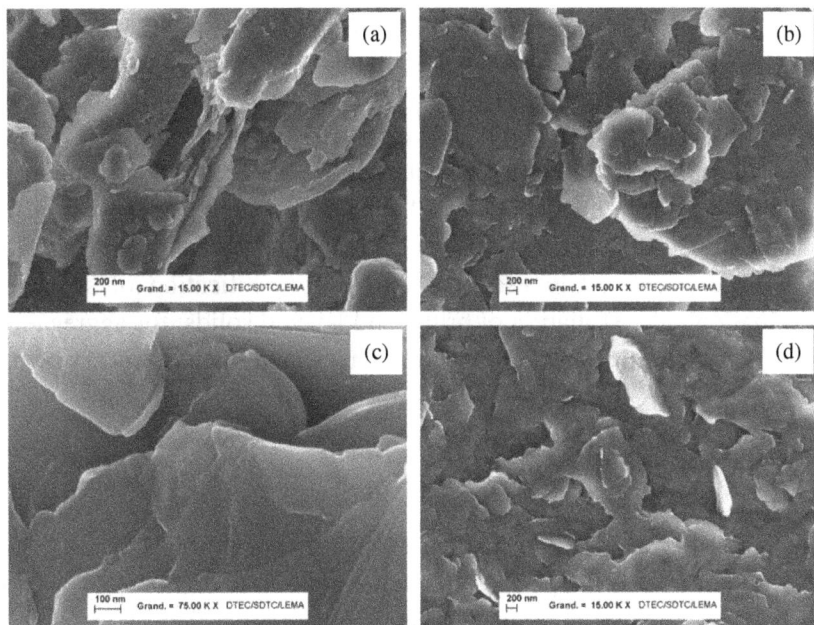

Figure 6: Scanning electron micrographs of vermiculite sonicated in H_2O_2 for (a) 1 hour, (b) 5 hours, (c) 7 hours and (d) 12 hours. Reprinted with permission from Ref. 52. Copyright (2016) Elsevier.

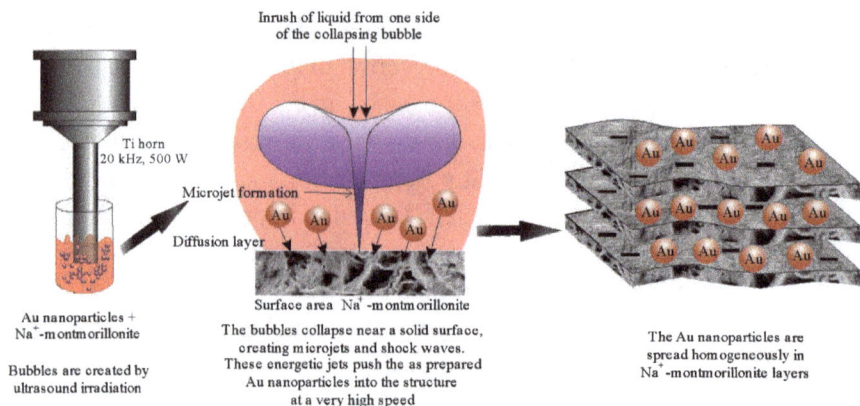

Inrush of liquid from one side
of the collapsing bubble

Ti horn
20 kHz, 500 W

Microjet formation

Diffusion layer

Surface area Na⁺-montmorillonite

Au nanoparticles +
Na⁺-montmorillonite

Bubbles are created by
ultrasound irradiation

The bubbles collapse near a solid surface,
creating microjets and shock waves.
These energetic jets push the as prepared
Au nanoparticles into the structure
at a very high speed

The Au nanoparticles are
spread homogeneously in
Na⁺-montmorillonite layers

Figure 7: Schematic illustration of intercalation of preformed gold nanoparticles into Na+-montmorillonite. Reprinted with permission from Ref. 53. Copyright (2016) American Chemical Society.

SEM, X-ray diffraction, and Fourier transform infrared measurements — strongly indicated that the sample loaded from 4.2 wt.% colloid solution is likely to be saturated by Au nanoparticles after 40 min sonication as optimum reaction time. Authors suggested that the microjets and shock waves provided by cavitation push the Au nanoparticles into the structure of clay (Figure 7). However, some fundamental studies are required to understand how low frequency ultrasound helps the intercalation processes.

To summarize, sonochemistry and especially physical effects of ultrasound have been widely used not only for modification of solid for delamination and exfoliation of brittle and layered solids, for aggregation of metal or ceramic powders but also for cleaning/depassivation of the surface of solids (Figure 2).

2.2. Synthesis of nanostructured solids

Sonochemical procedures were also widely developed for crystallization (reduction of nucleation periods and better control of crystal size) and for preparation of nanoparticles (more nanostructured particles, better distribution of size, better colloidal properties, etc.).

For example, mesoporous titanium dioxide nanocrystalline powders were synthesized by ultrasound-based hydrolysis reaction of tetrabutyl titanate in water without using any templates or surfactants that are necessary to add in conventional preparation.[54] Interestingly, the photocatalysts prepared under ultrasound (20 kHz, probe, 1,200 $W \cdot cm^{-2}$) led to better results in photocatalytic oxidation of mixtures of formaldehyde and acetone at room temperature compared to commercial photocatalysts.

Qiu *et al.* reported the synthesis of a microporous metal-organic framework $Cu_3(BTC)_2$ (where BTC is benzene-1,3,5-tricarboxylate) at room temperature and ambient pressure using ultrasonic irradiation (40 kHz, 60 W output power).[55] In terms of green chemistry, the advantages observed here were the high yields (62–85%) and short reaction times (only 5–60 min) compared with traditional thermal and solvothermal methods, as well as solvent diffusion techniques.

A one-step ultrasonic method (low frequency) was proposed for the synthesis of a carboxylate-functionalized multiwalled carbon nanotube-supported bimetallic platinum–cobalt nanoparticles catalyst.[56] The prepared catalyst showed excellent electrocatalytic activity in acid solution for oxygen reduction reaction and presented a high mass activity.

Recently, Seok *et al.* reported the sonochemical synthesis of palladium oxide (PdO) nanoparticles deposited on silica nanoparticles (Figure 8).[57] Through low frequency ultrasonic irradiations (20 kHz, 80 $W \cdot cm^{-2}$), the synthesis was performed at room temperature and atmospheric pressure with short reaction time (15 min). The prepared catalysts, with better surface areas, were efficiently used for the selective *s* oxidation reaction of alcohols in the presence of molecular oxygen.

Allahyari *et al.* studied the irradiation power and time for the co-precipitation of nanostructured $CuO–ZnO–Al_2O_3$ over HZSM-5 (zeolite).[58] See (Figure 9) Ultrasonic irradiation was performed under argon atmosphere using a probe at 20 kHz with different power outputs (50 W, 100 W and 150 W) and sonication times (30 min, 45 min and 60 min). The X-ray powder diffraction (XRD), field emission scanning electron microscopy (FESEM), energy-dispersive X-ray analysis (EDX), fourier transform infrared spectroscopy (FTIR) and BET analyses exhibited smaller particles with higher surface area and less population of particle aggregates at longer and highly irradiated nanocatalysts. For the longest irradiation time, the most intense power (150 W during 60 min) showed a very narrow particle size distribution. More than 65% of particles of this nanocatalyst were in the range of 1–30 nm.

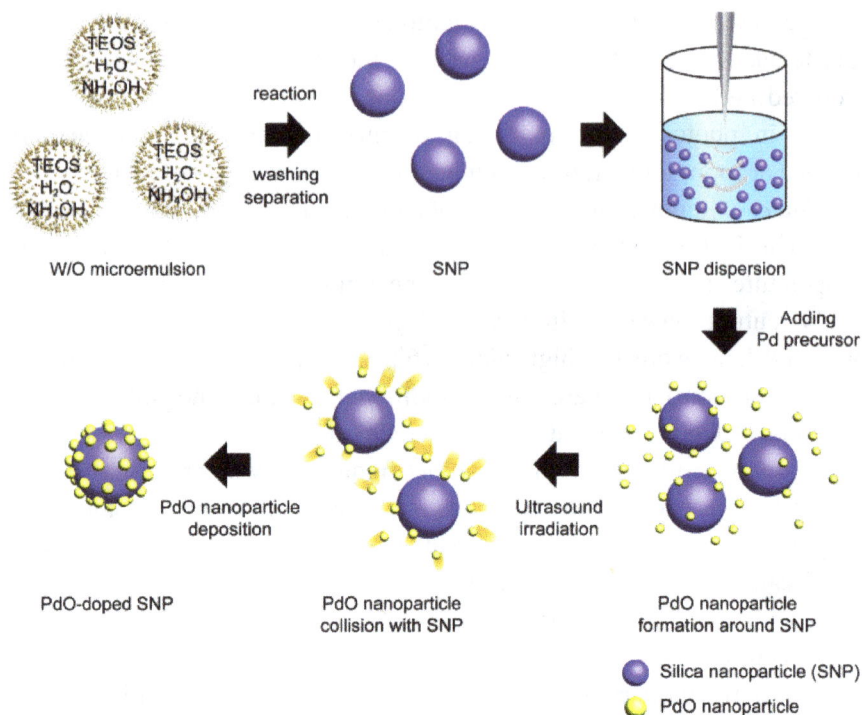

Figure 8: Procedure of synthesizing silica nanoparticles and PdO nanoparticules (TEOS: tetraethyl orthosilicate). Reprinted with permission from Ref. 57. Copyright (2016) Elsevier.

2.3. Contribution of physical and chemical effects of ultrasound

The most important physical effects of ultrasound arise from the high-speed jets and intense shock waves induced by cavitation. They are frequently used to prepare emulsions, agglomerate malleable materials, break down friable materials, modify solid surfaces and exfoliate layered materials. Enhanced mass transfer as a consequence of acoustic streaming and bulk thermal heating is another physical effect of high intensity ultrasound used in preparation of materials.[44] Physical effects of ultrasound can also be used to promote the diffusion of dopant ions into spherical nanoparticles.[59]

But under certain conditions, both the chemical and physical effects of ultrasound can play a synergistic role in the synthesis of materials.[44]

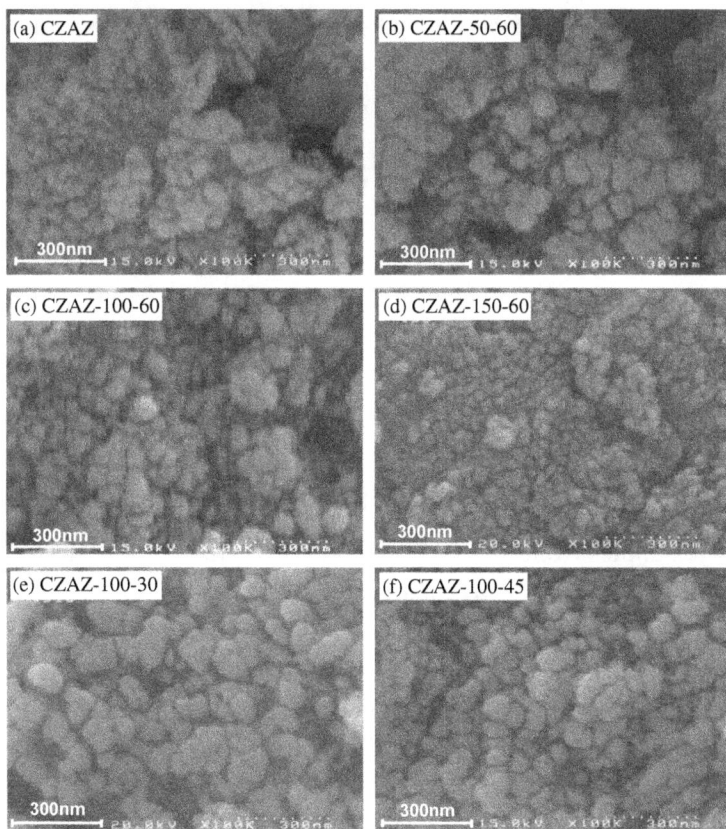

Figure 9: FESEM of synthesized nanocatalysts through ultrasound-assisted method with different powers (50 W, 100 W, 150 W) and times (30 min, 45 min and 60 min) and without ultrasound (a). Reprinted with permission from Ref. 58. Copyright (2016) Elsevier. Ref.

For example, a reactive solvent such as styrene can be sonochemically activated for the preparation of functionalized graphenes.[60] Indeed, polystyrene radicals can be formed and attack the surface of exfoliated graphenes to form functionalized graphenes. Some other examples were reported for the preparation of protein microspheres[61] or the preparation of organic latex beads,[62] showing the contribution of chemical effects of ultrasound.

From the green chemistry point of view, the use of ultrasound is interesting not only to reduce the time of preparation, amount of solvents and change the size, the distribution and structure of the particles, but also to provide to these catalytic materials new properties and reactivities, often showing better results than classically prepared catalysts.

3. Polymer Chemistry

In this section, we have chosen to discuss two utilizations of ultrasound in polymer chemistry. Indeed, sonochemistry cannot only be involved in the polymer formation, but also in the depolymerization or selective cleavage of polymers. The formation of polymer composites under ultrasonic activation will not be treated here.[63]

3.1. Ultrasound-based synthesis of polymers

Several examples reported sonochemical polymerization methods at lower temperatures and with enhanced rates, compared to more classical synthesis under silent conditions.[10,64] In general, the primary radicals that initiate the polymerization stem from solvent vapor cleavage in the cavitation bubbles. We note that monomers pyrolysis (for example, from styrene or methyl methacrylate)[65,66] can also occur in the cavitation bubble. Degradation of the polymer chains occurs simultaneously and high molecular weight polymers are formed at early stages of the reaction. Indeed, prolonged sonication time decreases the rate of conversion.

Interestingly, recent examples proposed the polymerization under heterogeneous conditions to form insoluble polymer products, by a gelation.[10,67] In this case, ultrasound drove sol–gel transition by favoring intermolecular hydrogen bonding between aggregates, which immobilized the solvent.

In the 1950s, emulsion polymerization under ultrasound was already proposed, even if it was really investigated in the 1990s.[10] This eco-friendly approach is based on the physical effects of cavitation, allowing an efficient micromixing and dispersion of the emulsion components and the formation of small oil droplet aggregation. Interestingly, no chemical initiators or stabilizers are required. The proposed mechanism is illustrated in Figure 10. In fact, primary radicals arising from water sonolysis (H$^\bullet$ and HO$^\bullet$) react with free monomer molecules, leading to monomeric

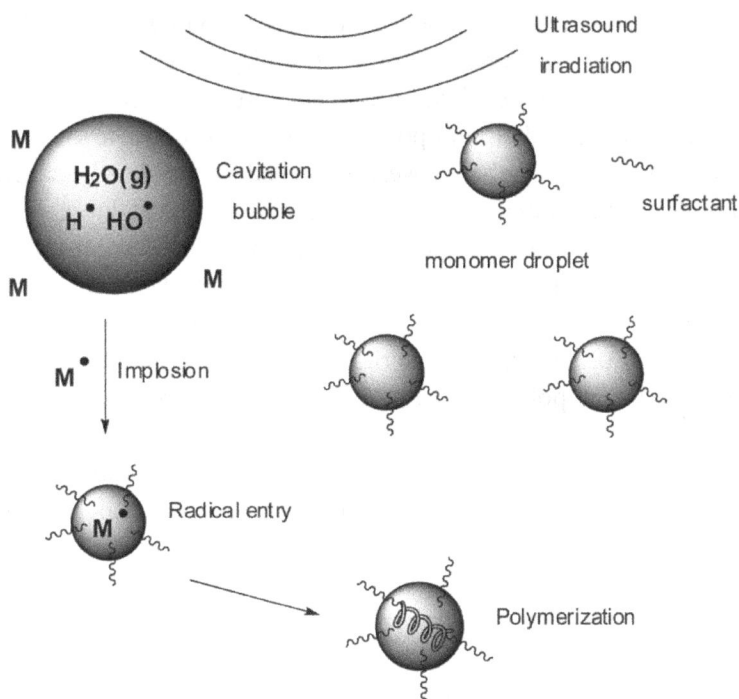

Figure 10: Representation of the emulsion polymerization under ultrasonic irradiation. Reprinted with permission from Ref. 10. Copyright (2016) Royal Society of Chemistry.

radicals in the emulsion system and initiating the polymerization. The unsaturated monomers are often volatile and may evaporate into the microbubble and be decomposed into hydrocarbon radicals (M•). The primary radicals produced are trapped by the solutes adsorbed at the bubble interface and release monomeric radicals in the bulk solution. Such radicals enter the monomer droplets and propagate radical polymerization. There has been a certain controversy regarding the role of surfactants.

3.2. Ultrasound-based cleavage and depolymerization of polymers

Cravotto and Cintas made a parallel comparison between sonochemical depolymerization and polymer mechanochemistry.[46] In fact, strong bonds may fragment during milling, grinding or shearing forming mechanoradicals with the concomitant decrease in polymer chain length. The type of cleavage is dominated by the stability of the resulting radicals which then undergo the usual pathways of addition and recombination.[68]

Based on the degradation of polymeric materials under low frequency ultrasound, two distinct phases were proposed in the depolymerization mechanism: (i) the initial scission in the midsection of large polymer chain followed by (ii) depolymerization of smaller molecular structures until the limiting molecular weight (M_{lim}) beyond which no further degradation is possible. Indeed, Moore *et al.* investigated at a fundamental level the degradation of polymeric materials under low frequency ultrasound and explained that polymer chain scission results from solvodynamic shear caused by cavitation.[69] In other words, a polymer molecule near the vicinity of a collapsing bubble is pulled toward the cavity of the bubble, and the solvodynamic shear elongates the polymer backbone, leading to scission. It was shown that scission generally occurs near the midpoint of a polymer chain (approximately within the middle 15% of the chain, in the case of homopolymers), where solvodynamic forces are the greatest see (Figure 11). The rate of cleavage from ultrasonic irradiation of polymer solutions depends on several experimental factors, including medium temperature, solvent, polymer concentration, initial molecular weight, ultrasound intensity, etc.

The **frequency** is probably the most influencial parameter. For example, Mohod and Gogate investigated the degradation of aqueous solution of polyvinyl alcohol and carboxymethyl cellulose under ultrasound at 20 kHz, 204 kHz and 694 kHz.[70] These experiments showed that the degradation was higher at 20 kHz compared to higher frequencies, due to better physical effects induced by low frequency ultrasound irradiation. However, Ulanski *et al.* reported that the depolymerization of starch in aqueous solution under high frequency ultrasound (360 kHz) was due to chemical effects of cavitation phenomenon.[71] Authors showed that the system became inefficient in the presence of a HO•-scavenger such as *tert*-butanol. Interestingly, the study showed that parameters such as ultrasonic power, polymer concentration and used atmosphere gas have an impact on the reaction yield. Depending on the chosen polymer and experimental conditions, the frequency can have a crucial role on the depolymerization or cleavage phenomena, even if no real trend can be drawn. Gogate and

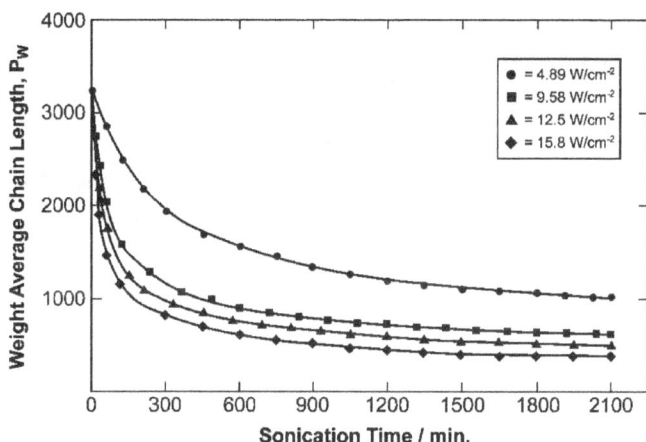

Figure 11: Effect of ultrasonic intensity on the chain scission of 25 mL samples of polystyrene in benzene (1 wt/vol %). Plot of chain length versus time in minutes for various ultrasonic intensities: (•) 4.89 W·cm^{-2}; (■) 9.58 W·cm^{-2}; (▲) 12.5 W·cm^{-2}; (◆) 15.8 W·cm^{-2}. Reprinted with permission from Ref. 69. Copyright (2016) American Chemical Society.

Prajapat suggested to combine low frequency ultrasound with oxidizing additives such as hydrogen peroxide in the presence of iron ions or ozone to enhance the generation of radicals to significantly degrade polymers.[72] Interestingly, the degradation yield can be controlled as a function of irradiation time or addition of radical scavengers.[73]

Gogate and Prajapat have recently reviewed all the literature on the subject and described the different factors affecting the polymer degradation[72]:

(1) **Ultrasound intensity:** higher intensities generally lead to a faster rate of chain scission with lower M_{lim}. For example, Mostafa reported that the weight average chain length (P_w) of a polystyrene solution in benzene (1 wt/vol%) decreased from 3,240 to 1,004 after sonication for 35 min with an intensity of 4.89 W·cm^{-2} while P_w decreased to 380 with a greater intensity (15.8 W·cm^{-2}).[74]

(2) **Solvent vapor pressure and viscosity:** the extent of degradation decreases with an increase in vapor pressure and viscosity of solvent.

(3) **Temperature:** the optimum temperature has to be determined since it directly affects the cavitation phenomenon and the generation of free

radicals. In addition, the effect of temperature is attributed to changes in the physical state of the system (vapor pressure, viscosity of the solvent, etc.).

(4) **Solution concentration:** low concentration and low volumes are favorable for the scission of polymers. At higher concentrations, the cavitation phenomenon can be reduced due to the more important viscosity, or the need of greater power density.

(5) **Initial polymer molecular weight and the chemical structure of the polymer:** it was reported that ultrasonic chain scission proceeds more rapidly at higher molecular weights. In addition, the substituent or the functional groups present on the polymer chain can also play a significant role in the ultrasonic depolymerization process.[72]

Recently, Simon's[75] and Moore's[76] groups worked simultaneously on the role of mass and length in the sonochemical cleavage of polymers. Two different polymer backbones were used (poly(styrene) and poly-(norbornene imide alkyne)) and irradiated at 20 kHz (10.4 W·cm^{-2}).[75] Experimental data and calculations showed that the molecular weight reduction upon sonication is the same for polymers with the same contour length.

(6) **Selection of operating and geometric parameters of sonochemical reactors:** the designing of reactors can clearly maximize the depolymerization process under ultrasound.

4. Lignocellulosic Biomass Conversion

Conversion of lignocellulosic biomass is a challenging topic of research in academic and industrial areas.[77,78] Indeed, this lignocellulosic biomass, essentially composed of cellulose, hemicellulose and lignin polymeric chains (Figure 12), represents the world's most abundant renewable carbon reservoir as a potential feedstock for producing chemical,[79–82] fuels[83–85] and materials.[86,87] In addition, among the large variety of biomass feedstocks available on Earth, lignocellulosic biomass does not compete with human food, an ethical problem often encountered with other renewable resources such as starch, vegetable oils, etc.[88]

Interestingly, ultrasonic energy can meet challenges of processing recalcitrant, multicomponent and heterogeneous lignocellulosic biomass.

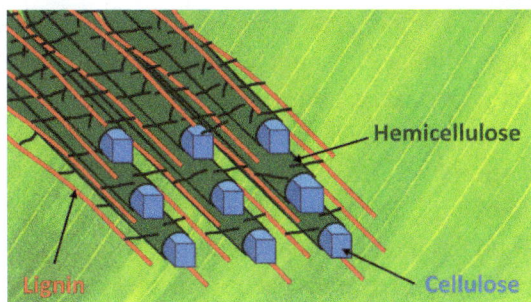

Figure 12: Schematic representation of the subcomponents in lignocellulosic biomass.

Indeed, ultrasound provides a severe physico-chemical environment that is difficult to obtain with other engineering methods. After treatment, no remarkable change in the chemical structure of biomass and in mechanism reactions is observed,[89] but reaction kinetics are generally accelerated, and therefore the efficiency and economics of the biomass conversion process are enhanced.[90,91] For this last reason, the introduction of ultrasonic energy plays a positive influence essentially on the biofuels production or for the pretreatment and extraction of biomass.[92,93] Often, the possibility to produce high-value chemicals is not identified or reported. Here, we have reported some interesting examples on the pretreatment of biomass for fuels production, extraction processes (see Section 5, page 95) and routes to produce chemicals of interest, involving ultrasonic technology.

4.1. Sonochemical pretreatment

A recent review proposed by Bussemaker and Zhang reported all the effects of ultrasound on lignocellulosic biomass as a pretreatment method.[94] The efficacy of ultrasonic pretreatment for improved processing was attributed partly to physical effects. Indeed, ultrasound improves mass transfer (enhanced turbulence, microlevel mixing) and enhances the accessibility of the biomass for subsequent processing. In this section, the goal is not to provide a complete list of ultrasonic pretreatment examples from the literature, but rather highlighting some interesting approaches.

Generally, the consequences of the use of ultrasound on lignocellulosic biomass are the increase of lignin extraction yields[94–97] and the

85

degradation of lignin components, especially through the cleavage of the lignin-hemicellulose linkages. Sonochemical effects can also be at the origin of the depolymerization and degradation of cellulose and hemicellulose *via* a cumulative effect of the hydroxyl radicals, shear forces and pyrolytic degradation of hydrophobic polymers in the hot regions around the collapsing bubbles.[98] Some practical considerations are essential for the use of ultrasound as a pretreatment option.[94] Indeed, some experimental attention is important to take into account factors such as the feedstock (type of biomass, loading), the solvent (pretreatment), the ultrasonic environment (gas, medium temperature), the reaction time (kinetics), the reactor configuration, the involved energy, the process cost and the scale-up.[94] Interestingly, some of these considerations are more general and concern all the biomass processes. These parameters will be discussed for conversion into chemicals later in this chapter.

It is also important to highlight that some examples also reported an inefficiency of ultrasound in pretreatment methods. For example, the chemical pretreatments (acid or basic) of soybean fiber showed significant lignin degradation, but ultrasonic pretreatment (20 kHz, maximum power output 2,200 W) was inefficient.[99] Authors attributed this to the fact that their starting material already had complete disruption of cellular structure of the extruded soybean flakes and ultrasound did not contribute to further disintegration of the cellular matrix. In this case, another explanation could be that good sonochemical conditions had not been determined.

Some reviews[94,100,101] and recent publications[102–106] reported the advantageous results provided by sonochemical pretreatment, essentially at low frequencies (20–50 kHz) obtained through the physical effects of cavitation.

Interestingly, at low frequencies and in considered experimental conditions, the pretreatment of cellulosic materials causes some hydrolytic changes (chain cleavage) and decrease of the Molecular Weight (M_w) but globally, did not show oxidative functionalization of materials.[89,107] However, the increase of the sonication time can yield a high crystallinity index of cellulose.[108]

Some authors paid particular attention to the investigation of the different sonochemical parameters. For example some studies reported the comparison between the effects of low and high frequencies. Yasuda *et al.* demonstrated better enzymatic hydrolysis of cellulose from unbleached Kraft pulp at 20 kHz and 28 kHz rather than at 500 kHz.[109] Through a systematic and rigorous analysis of parameters, Bussemaker *et al.* reported the improvement of wheat straw fractionation at 40 kHz, 376 kHz and 995 kHz, *via* different mechanisms. In fact, at low frequency, delignification was favored by the ultrasonic treatment, while at higher frequency, carbohydrate solubilization was preferred.[110] The particle size and loading of biomass also have an importance in the results. This study clearly shows the importance of the sonochemical parameters control to better understand the ultrasonic effect on pretreatment of biomass, and in a more general way, in sonochemistry.

Even if the first goal of pretreatment is to make the starting feedstocks more accessible, some chemicals such as hexose, pentose or glucose can also be obtained, with high yield under low frequency ultrasound.[111–113] For example, the pretreatment of microcrystalline cellulose by 20 kHz ultrasonic irradiations was reported, allowing to increase the overall yield of glucose up to 42 wt.% in the presence of a solid acid catalyst.[112]

Subhedar and Gogate established the superiority of ultrasound-assisted pretreatment, compared to the chemical, physical, physicochemical and biological pretreatments to improve the hydrolysis process *via* the reduction of the structural rigidity of lignocellulose and the elimination of the mass transfer resistances.[114] For authors, the major advantages to use ultrasound are the reduction of processing time for similar results, the operation at lower temperature and the need of lower amounts of reactants (i.e., enzyme). They think that this technology is an "add-on" step to the existing process, and they suggested that ultrasound-assisted process can also be coupled with other conventional techniques, such as alkaline pretreatment, as a promising technique for efficient bioethanol production.

4.2. Sonochemical production of chemicals

We have reviewed the literature about the use of ultrasound to produce chemicals from lignocellulosic biomass.[115] Indeed, the ultrasonic technology is mainly used for low-value chemical production such as biodiesel. Some preliminary studies demonstrated that the access to added-value chemicals can be easily and sometimes solely obtained by the use of ultrasound. In addition, the design of sonochemical parameters offers many opportunities to develop new eco-friendly and efficient processes.

The production of chemicals through enzymatic sonochemical reactions will be developed in Section 6.4 of this chapter (page 123). For more information, we refer to the publications proposed by Bremner *et al.*[116] and Yachmenev *et al.*[117] This subsection is rather focused on organic sonochemistry from biomass substrates to access to more complex compounds.

In 2001, Kardos and Luche proposed a complete review on sonochemistry of carbohydrate compounds, reporting the use of ultrasound for depolymerization of starch, dextrans or other polysaccharide derivatives, but also several reactions from sugars such as hydrolysis, oxidation, acetalization and glycosylation.[98] Authors particularly reported the importance of energy, temperature, frequency and ultrasonic equipment in the obtained results. Globally, even if some mechanisms remain unknown, sonication displays a beneficial role on the reaction rates, yields, and chemo-, regio- and stereoselectivities for common reactions from carbohydrates derivatives.[98]

Some reactions assisted by ultrasound are reported in the literature from different biomass feedstocks (Table 2). Ragauskas *et al.* generated condensed phenols during ultrasonic irradiation of softwood Kraft lignin, providing high molecular weight lignin. This ultrasonic process is competitive with recent enzymatic and non-enzymatic processes, and proposes a simple experimental design and fast polymerization rates (Table 2, entry 1).[118] Applied on highly purified Kraft lignin in aqueous alkaline solution, this ultrasonic-assisted method resulted in the formation of a high molecular mass fraction (35% of lignin content) with an average molecular weight 450-fold greater than the original sample after less than 1 hour of treatment. In addition, ^{13}C NMR and ^{31}P NMR analyses indicated that the polymerized fraction was enriched with C_5 condensed phenolic structures. Napoly *et al.* investigated the oxidative depolymerization of Kraft lignin in the presence of H_2O_2 and different metal salt catalysts in

Table 2: Catalytic ultrasound-assisted reactions for chemicals production from biomass feedstocks.

Entry	Biomass origin	Ultrasonic conditions	Experimental conditions	Main observations	Reactions	Reference
1	Kraft lignin	Frequency not indicated (probably low frequencies), output amplitude 35% of 500 W, probe	60 min, 15°C, NaOH aqueous solution (pH = 12)	Formation of high molecular mass fraction with a weight average molecular weight over 450-fold greater than the original sample	Polymerization	118
2	Kraft lignin	20 kHz, 0.38 W·mL⁻¹ as acoustic power, probe	H_2O_2 (35% aqueous solution), 60 min, 45°C and 80°C, $Na_2WO_4 \bullet 2H_2O$	Recombination of formed fragments under ultrasound. Changes in the lignin structures	Oxidative depolymerization	119
3	Municipal wood waste	24 kHz, power output 85 W·cm⁻²; probe	2 hours, 160–180°C, p-toluenesulfonic acid, sulfuric acid (3% w/w), diethylglycol	The reaction times were shortened by a factor of up to nine, when using the ultrasound process, yielding smaller residual particles. No influence on the hydroxyl number of the final products	Liquefaction process	122
4	Pulp and paper kraft mill (from aspen and poplar processing)	357 kHz, 2.4 W·cm⁻² and 4.0 W·cm⁻²	25 hours, 20°C, $K_2Cr_2O_7$, H_2SO_4	Improvement in bleaching of chromophores and reduction in turbidity	Total degradation	123
5	α-D-glucoside and sucrose	20 kHz and 500 kHz, acoustic powers: 0.01–0.6 W·mL⁻¹, probe	5°C, 0.65 mol% TEMPO, NaBr, 12% aqueous NaOCl, pH = 10.5	Increase of the TEMPO-mediated oxidation rate of methyl glucoside/sucrose. No need of NaBr under ultrasound	Oxidation reaction	124–126

(Continued)

89

Table 2: *(Continued)*

Entry	Biomass origin	Ultrasonic conditions	Experimental conditions	Main observations	Reactions	Reference
6	Cotton linter pulp	40 kHz, 300 W, 10 L ultrasonic bath	25°C, TEMPO, NaBr, NaOCl, pH = 10,	Direct production of cellulose nanocrystals with high carboxylate content. Change of cellulose crystallinity during TEMPO-assisted oxidation	Oxidation reaction	127
7	Mill-bleached, machine-dried hardwood kraft pulp	68 kHz and 170 kHz, 1,000 W, ultrasonic bath	90 min, 25°C, 4-acetamido-TEMPO, NaBr, NaOCl, pH = 10.0–10.2	Reaction more efficient and faster under ultrasound at a given temperature	Oxidation reaction	128
8	Bleached hardwood kraft pulp	24 kHz and 170 kHz, 500 W·L⁻¹ and 262 W·L⁻¹, bath and sonoreactor	4-acetamino-TEMPO, NaBr, NaOCl.	Decrease of energy consumption with higher COOH production rate under ultrasound	Oxidation reaction	129
9	D-Glucose	35 kHz, ultrasonic bath	5 min, RT, 1wt% Au/SiO₂ 30% H₂O₂, pH = 9.2	Sonication allowed the catalyst to be stably suspended, providing high conversion, high selectivity and high reproducibility	Oxidation reaction	130
10	D-fructose	20 kHz, 26 W·cm⁻², 78 W·cm⁻² and 130 W·cm⁻², probe in a laboratory-scale pressure autoclave.	70°C, 90°C or 110°C, 3 tested catalysts: Cu/SiO₂, Raney-Ni, CuO/ZnO/Al₂O₃, H₂ (10 bar, 30 bar and 50 bar)	Influence of ultrasound depends on the selected catalyst. No influence of pressure on sonication effect	Hydrogenation	131

acetone/water mixture at low temperature under low frequency ultrasound activation (20 kHz, Table 2, entry 2).[119] In these experiments, the production of monomers decreased under ultrasound compared to silent conditions, due to the recombination of formed fragments. However, some changes were reported on the lignin structure, with the increase of C–O–C bonds and a modest oxidation of aliphatic –OH of side chains. The authors showed that ultrasound coupled with hydrogen peroxide induced stronger oxidative conditions, involving higher oxidative coupling of phenoxy radicals issuing from lignin polymerization.[119]

These two examples show that sonochemistry can be a new potential process for the challenging valorization of lignin in high-value added chemicals production, especially for higher high-molecular-weight applications.[120]

In a recent contribution, the use of TiO_2 particles in combination with 24 kHz frequency ultrasonic irradiation was reported for lignin degradation in pretreatment of kenaf biomass.[121] In addition to this sonocatalytic reaction, the Fenton reaction based on Fe^{2+} cations as well as the combined reactions were also studied. The delignification effect was due to the synergistic sonocatalytic-Fenton reaction that was shown to enhance the production of HO• radicals and hence the lignin conversion. In this case, the production of monomers was not investigated.

Kunaver *et al.* chose to treat common wood waste materials in order to provide an alternative strategy for the recycling of waste that occurs in municipal waste deposits (Table 2, entry 3).[122] Under 24 kHz ultrasonic irradiation and in the presence of acids, the reaction times were shortened by a factor of up to nine, compared to silent conditions (10 min versus 90 min), owing to smaller residual particles. The addition of diethylene glycol during the process allows the reduction of the medium viscosity. Authors showed the impact of viscosity on the reaction, by testing several ultrasonic intensities (85 $W \cdot cm^{-2}$, 51 $W \cdot cm^{-2}$ and 17 $W \cdot cm^{-2}$). At higher viscosities, cavitation is more difficult to induce (a greater power input is required) and the number of cavitating bubbles per unit volume is reduced. Interestingly, authors determined the process energy input of 0.44 $kW \cdot kg^{-1}$ and 1.35 $kW \cdot kg^{-1}$ in the presence of ultrasound or not, respectively.

Shaw and Lee showed that high frequency ultrasound irradiation (357 kHz) of pulp and paper final effluent in the presence of concentrated

OH
HO
HO
OH
O
OMe
TEMPO/NaBr/NaOCl
20 or 500 kHz ultrasound
ONa
O
HO
HO
OH
O
OMe

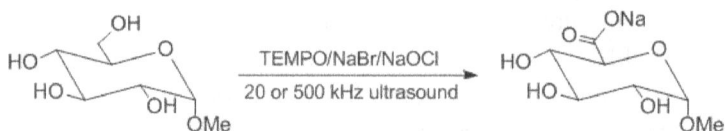

Scheme 11: Reaction scheme of the oxidation of methyl-α-D-glucoside by TEMPO-system.[124]

sulfuric acid and potassium dichromate results in the bleaching of chromophores and a reduction in turbidity (Table 2, entry 4).[123] As no decrease in chemical oxygen demand (COD) was observed, authors suggested that the hydroxyl radicals were scavenged by bicarbonate and sulfate ions present in the final effluent. These radical scavengers can represent some limitations by reducing the efficiency of radical mechanisms.

In the 2000s, Queneau *et al.* investigated the TEMPO-mediated oxidation of glucosides and sucrose in the presence of sodium hypochlorite in basic aqueous solution (Scheme 11, Table 2, entry 5).[124–126] Through several studies of frequencies (20 and 500 kHz) and experimental conditions, they attempted to explain the reaction mechanisms, showing the formation of the nitrosonium ion (NO$^+$, the active oxidizing species in the catalytic cycle) potentially formed *via* homolytic cleavage of chlorine or bromine or their hypohalous acids. In these studies, authors tried to demonstrate, separately, the sonophysical and sonochemical aspects of ultrasound in the field of carbohydrate chemistry, through the control of the frequency parameter.

The TEMPO/NaBr/NaClO oxidizing system was also used by Qin *et al.* from cotton linter pulp under 40 kHz ultrasound (Table 2, entry 6).[127] In this example, they reported the direct production of cellulose nanocrystals having high carboxylate content (1.66 mmol$\cdot$g^{-1}) which was stable in water. During oxidizing reaction, some of the amorphous region in the cellulose fiber (cotton linter pulp) was modified and gradually hydrolyzed. However, the crystalline part of cellulose fiber was not modified. Interestingly, the coupling between ultrasound and TEMPO/NaBr/NaClO chemical system led to a slight increase of cellulose crystallinity and allowed a reduction in time of treatment.

High-intensity ultrasound had also been used in combination with a TEMPO system to oxidize the primary hydroxyl groups on the cellulose polymer chains to carboxylate groups at 68 kHz and 170 kHz (Table 2, entry 7).[128] An increase of 10–15% in carboxyl content (1,071 mmol·kg^{-1} versus. 938 mmol·g^{-1}) and in nanocellulose yield (+10%) was observed under optimized conditions. This research group also mentioned that reaction temperature, medium pH and frequency significantly affected the properties of oxidized pulp. For example, the ultrasound treatment at 68 kHz gave lower yield than that at 170 kHz.[128]

To reduce the energy consumption of the process, Paquin *et al.* proposed the transfer from a batch mode to a continuous flow-through sonoreactor for cellulose oxidation *via* TEMPO-mediated system (Table 2, entry 8).[129] The flow-through reactor increases the reaction rate by 36% compared with the glass reactor in batch mode, decreasing the energy consumption by 87.5% in term of nominal input power. The efficiency of this scale-up was demonstrated by potassium iodide dosimetry indicating the possibility of scaling up such reactions to an industrial scale in continuous mode.

Bujak *et al.* provided a catalytic system in aqueous solution at room temperature for D-glucose oxidation with high conversion and selectivity (Table 2, entry 9).[130] The use of low frequency ultrasound allowed the Au/SiO$_2$ catalyst to be stably suspended and provided a high reproducibility, that can sometimes be problematic with the use of an ultrasonic bath. This example remains interesting, since the combination of ultrasound technology with the eco-friendly oxidant, aqueous hydrogen peroxide (H_2O_2), led to excellent conversion and selectivity. This example on D-glucose oxidation based on the H_2O_2/ultrasound combination (Table 2, entry 9) is promising for future investigations on real biomass samples.[130]

As shown in Table 2, the main reaction studied under ultrasound is oxidation of biomass feedstocks in combination with an oxidant. This can be explained by the chemical effects (production of radicals) provided by ultrasound. However, low frequency ultrasound is mainly used, probably because of their best commercial availability. In this context, the investigation of high frequency is clearly necessary to understand how radical oxidation can be used for biomass valorization.

An interesting example of hydrogenation is also reported in the literature (Table 2, entry 10).[131] Indeed, the hydrogenation of D-fructose into D-mannitol was investigated under different pressures of dihydrogen (10 bar, 30 bar and 50 bar) and using different catalysts (Cu/SiO$_2$,

Raney-Ni and $CuO/ZnO/Al_2O_3$). The choice of the catalyst had a significant influence on ultrasound effects. With the $Cu/ZnO/Al_2O_3$ catalyst,
ultrasound moderately decreased the reaction rate. In the case of Raney-Ni
catalyst, a slight increase in the reaction rate was observed under sonication. In fact, the essential improvement of the catalyst activity under
acoustic irradiation was observed using Cu/SiO_2. However, selectivity
remained unaffected by sonication in this reaction. With the best catalyst
(Cu/SiO_2), authors studied the optimal conditions in terms of temperatures
(90°C and 110°C), H_2 pressure (50 bar) and ultrasonic power input
(30 W). What is remarkable in this study is the use of a laboratory-scale
pressure autoclave with possibility of ultrasonic irradiation (Figure 13).

All these examples clearly show that the use of ultrasound in biomass
valorization is in its infancy, especially for the production of chemicals and
intermediates of interest. However, even if the chemistry is not yet mastered for the conversion of biomass into chemicals, the preliminary and
existing results are promising, and several tracks have already been opened.
In most cases, the functionalization of biopolymers is favored. For example, ultrasonic treatment of potato, wheat, corn and rice starches in water
or ethanol improved their depolymerization leading to the higher abilities

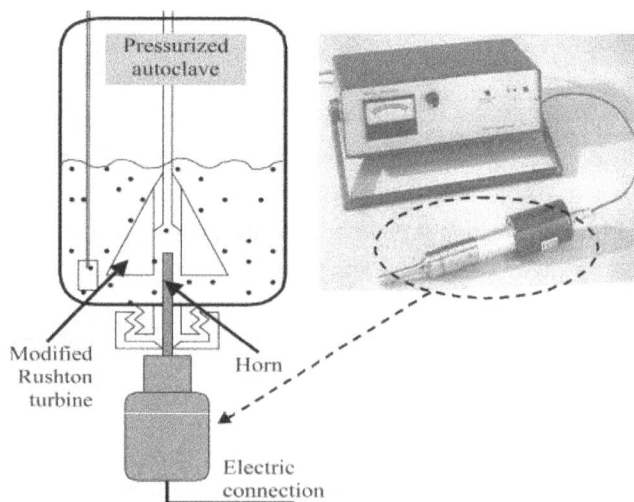

Figure 13:   High-pressure autoclave with *in situ* ultrasonic irradiation system. Reprinted
with permission from Ref. 131. Copyright (2016) American Chemical Society.

of fat and water absorption, lower viscosities and higher least gelling concentration (LGC), solubility and swelling power in comparison to their native counterparts.[132,133] In addition, potato starch showed higher paste clarity after sonication, which is desirable in many food applications.

At this stage of research in the area, there are as many results as reported examples. Indeed, all cases are different in the function of the choice of biomass feedstocks and the ultrasonic parameters that authors used, exactly known or not. It is really difficult to identify clear trends, except the improvements in terms of reaction time reduction, yields and mass transfer effects, and the huge potential that sonochemistry can bring in the area. Along with low frequency ultrasound irradiation, the investigation of high frequencies (>150 kHz) is also encouraged to benefit from the chemical effects of the ultrasonic irradiation for oxidation reactions. In conclusion, the real challenge is to understand how ultrasound can be beneficial during biomass conversion processing, and what the most promising ways to explore are.

5. Sonochemical Extraction

Sonochemical extraction, often called UAE in the literature (*UAE* for *Ultrasound-Assisted Extraction*) is becoming a real clean and green extraction technology. Indeed, the use of ultrasound in this area allows in many cases to (1) enhance extraction yield; (2) enhance aqueous extraction processes without organic solvents; (3) provide the opportunity to use alternative and eco-compatible solvents by improving their extraction performance and (4) enhance extraction of heat-sensitive components under conditions that would otherwise have low or unacceptable yield.[134] The use of ultrasound is an attractive extraction technology for various molecules and biomaterials,[135] including polysaccharides,[136] essential oils,[137] proteins and peptides,[138] fine chemicals (dyes and pigments),[139] bioactive molecules of commercial importance[140] and metals.[141,142]

In the lignocellulosic biomass area, the direct extraction of sugars, hemicelluloses or lignin from the raw biomass was reported. Table 3 presents some examples from different biomass feedstocks, with different experimental and ultrasonic conditions and the main associated observations on the role of ultrasound in the extraction process. As a matter of fact,

Table 3: Some examples of ultrasound-assisted extraction from lignocellulosic biomass feedstocks.

Entry	Biomass origin	Ultrasonic conditions	Experimental conditions	Main observations	Reference
1	Coconut shell powder	25 kHz, 150 W (ultrasonic bath)	60 min, 100°C, ethanol/water	Decrease of extraction time	150
2	Longan fruit pericarp	40 kHz, 120–241 W (ultrasonic bath)	18–22 min, 51–70°C, water	Decrease of extraction time, Changes in the microstructure.	147
3	Bamboo	35 kHz, 100 W (ultrasonic cell crusher)	5 min, 20°C, ethanol	Slight increase of the molecular weight of the lignin; increase of thermal stability of the prepared fractions	143
4	Prunella vulgaris	40 kHz, 200 W (ultrasonic bath)	30 min, 40–80°C, ethanol/water	Enhancement of flavonoids extraction yield	152
5	Poplar wood	20–24 kHz, 570 W (horn)	30 min, 25°C, ethanol, methanol and dioxane	75% of the original hemicelluloses extracted	153
6	Sugarcane bagasse	Not indicated	45 min, 55°C, alkaline peroxide	Mechanical effects	144
7	Defatted soy flakes	20 kHz, 0.30–2.56 W·mL^{-1} (booster + horn)	15–120 s, 4°C, water	Improvement of protein and sugar extraction	145
8	Cassava chips	20 kHz, 2–8.5 W·mL^{-1} (horn + booster)	10–30 s, 0.05 M acetate buffer	Reaction time reduced by 24 hours; Increase of ethanol yields	151
9	Dewaxed wheat straw	20 kHz, 100 W (horn + booster)	5–35 min, 60°C, 0.5 M NaOH methanol/water	Increase of hemicelluloses yields by 9.2%	154
10	Grape pomace	37 kHz, 140 W, 20°C (ultrasonic bath)	2.6 hours, 20°C, 0.4 m KOH, water	Reduction of extraction time, use of lower concentrations of KOH.	146

only low frequencies (20–40 kHz) are used since mechanical effects are expected, even if in some publications, authors admitted that they do not have any precision on the mechanism of ultrasound in their extraction and that further studies are needed.[147–149] The solvents used during the extraction process (polarity, mixing, pH, etc.) can also be a key parameter (Table 3). Globally, it was reported that ultrasound decreased the extraction time (between few seconds to hour),[150,151] improved extraction yields,[152–154] even if optimal ultrasonic conditions often need to be investigated more deeply.

Chemat *et al.* reported some examples of large-scale ultrasound-assisted extractions as tools for biorefinery concept.[155] For example, the extraction of polyphenols (phenolic acid derivatives and flavonoids) was performed at a pilot-scale (30 L) in water under ultrasound irradiation (25 kHz, 0.764 W $\cdot$ cm^{-2}, 40 min at 40°C) from apple pomace (solid wastes from apple juice or cider production).[156] In this case, a 15% increase in total polyphenol content in the extract was obtained compared to extraction by maceration (Figure 14).

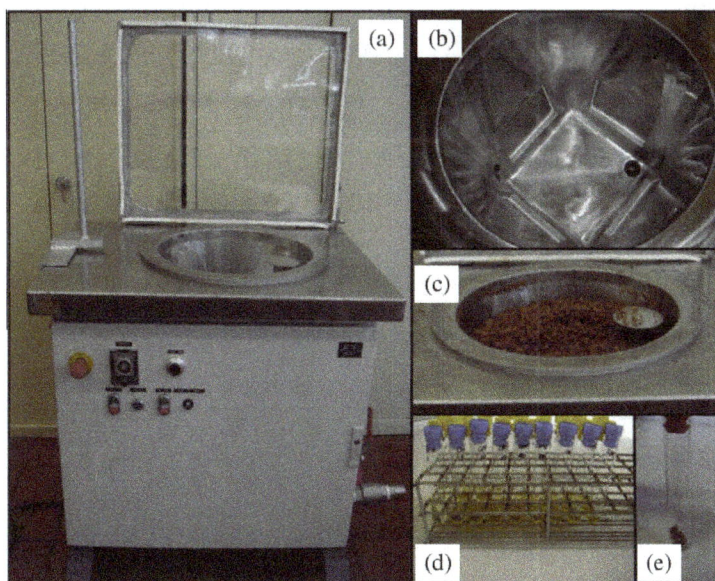

Figure 14: (a) Large-scale ultrasound apparatus; (b) inside view of the apparatus; (c) UAE of apple pomace; (d) liquid extract and (e) lyophilized extract. Reprinted with permission from Ref. 156. Copyright (2016) Elsevier.

The mechanism of UAE using solvent and solid matrix is based on (1) an enhanced mass transfer *via* turbulent mixing and acoustic streaming; (2) a surface damage at solvent–matrix interfaces by shock waves and microjets; (3) a high-velocity interparticle collision and (4) a disintegration of matrix to increase surface area.[134]

In conclusion, UAE processes have proven to be a green technology in different areas such as food, environmental, pharmaceutical and analytical chemistry: extraction intensification in aqueous media, low energy consumption and enhanced extract quality were reported compared to conventional maceration.[155] Indeed, the majority of classical methods of extraction involve corrosive or toxic chemicals (such as chlorinated solvents) and long extraction times. Interestingly, Shirsath *et al.* suggested the combination of ultrasound with supercritical fluid extraction as a promising technique to provide a significant degree of intensification.[135]

6. New Combinations ... Towards New Results!

The combination of ultrasound with other innovative technologies starts to bring some synergetic effects. The first reported examples of ultrasound/ionic liquids, ultrasound/microwave, ultrasound/electrochemistry and ultrasound/enzymes couplings showed interesting improvements and represent a great potential in terms of innovation. In these cases, it will be important to separately compare silent, coupled and non-coupled conditions to highlight the synergetic effects of the combination.

6.1. Ultrasound and ionic liquids combination

Ionic liquid (IL) media as well as sonochemistry are two recently developing technologies used in different fields of chemistry. Often, their uses in a reaction or process produce improvements in terms of efficiency, selectivity, yield, reaction time, recycling and/or, in some cases, other unexpected results. The idea of combining these two efficient technologies was innovative and produced clear synergies in some cases, but it is important to look closely at the literature to understand how this unique combination can be optimized as a highly effective approach to a number of chemical processes.[157] We will particularly develop in this subsection the goal to not only show the synergetic effects brought by this combination in different chemical fields presented previously in this chapter, but also to describe the research approaches developed to better understand how it works at a more fundamental level.

In the last two decades, the use of ILs as reaction media reported in different areas of chemistry has considerably increased.[158,159] These solvents, commonly defined as a class of salts with low melting points (typically less than 100°C)[160] or no melting point, have attracted intense attention from chemists. Composed of a bulky organic cation associated to an anion, organic or not, ILs present some unique properties often unavailable with traditional solvents, including, in various cases, negligible vapor pressure, high chemical and electrochemical stabilities, high polarity, etc.[160–163] New approaches involving ILs have been proposed for aspects of energy chemistry,[164,165] preparation of materials,[166] biomass valorization,[167,168] analytical chemistry,[169] microextraction,[170] organic and pharmaceutical chemistry,[171] electrochemistry[172] and many others. Interestingly, we estimate that nearly one million simple ILs could be easily prepared in the laboratory, leading to more than 10^{18} possible ILs.[173] This represents a great advantage not only in terms of tunability for these solvents, but also a limiting drawback in terms of the lack of theoretical and fundamental data about their properties to guide their use.

From our literature survey, we highlighted the main applications involving the ILs/ultrasound combination: (1) use of ultrasound for the synthesis of ILs; (2) synergetic effects found in organic chemistry; (3) for materials preparation (catalysts, nanoparticles, nanotubes, etc.); (4) for extraction and microextraction; (5) for biomass processing such as pretreatment of lignocellulosic feedstocks and (6) others applications (electrochemistry, sonochemical degradation of ILs, etc).[157]

6.1.1 Applications in organic chemistry

The first example of this combination was reported by Srinivasan *et al.* in 2001 to promote C–C bond formation *via* an Heck reaction under ambient conditions in 1,3-di-*n*-butylimidazolium bromide ([C_4C_4im]Br) and 1,3-di-*n*-butylimidazolium tetrafluoroborate ([C_4C_4im][BF_4]) with an ultrasonic cleaning bath (50 kHz).[174] The ultrasound-assisted Heck reaction of the

Table 4: Heck reaction of iodobenzenes with activated alkenes under sonication in $[C_4C_4im]Br/[C_4C_4im][BF_4]$ ILs.[174]

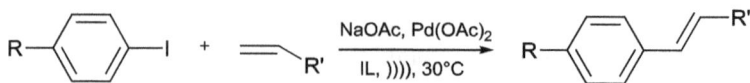

Entry	R	R′	Sonication time (hours)	Isolated Yield (%)
1	–H	–COOMe	2.0	81
2	–H	–COOEt	1.5	87
3	–H	–Ph	1.5	82
4	–OMe	–COOMe	3.0	82
5	–OMe	–COOEt	3.0	79
6	–OMe	–Ph	3.0	80
7	–Cl	–COOMe	1.5	79
8	–Cl	–COOEt	1.5	77
9	–Cl	–Ph	1.5	73

iodobenzenes with alkenes at 30°C showed complete conversion in just 1.5–3 hours to afford the desired products in excellent isolated yields (73–87%, Table 4). No reaction under similar ultrasonic conditions was observed when the ILs were replaced by molecular solvents such as dimethylformanide (DMF) and *N*-methyl-2-pyrolidine (NMP) and no reaction, even in traces, was observed under ambient conditions in the absence of ultrasound. Here, there is a clear, combined effect, confirming that both IL and ultrasound are required for the reaction. In addition, under these conditions, the reaction rates were considerably enhanced and reaction times decreased, compared to the classical conditions. The authors demonstrated *via* nuclear magnetic resonance (NMR/MS) and *in situ* TEM analyses the formation of a Pd–biscarben complex and its subsequent sonolytic conversion to a highly stabilized cluster of zero-valent Pd nanoparticles; this suggested that the enhancement in rate and yields were, at least in part, due to the formation of a new, highly active catalytic species.

From this first example, the same research group studied other organic reactions such as Suzuki cross-coupling,[175] the nitration of phenols,[176] the acetylation of alcohols,[177] the synthesis of 3,4-dihydropyrimidin-2-(1*H*)-ones,[178] the Sonogashira reaction[179] and the synthesis of 1,8-dioxo-octahydro-xanthene derivatives.[180] Interestingly, as all of these experiments are performed by the same research group, the ultrasonic conditions are similar (cleaning bath, 50 kHz, same acoustic power), thus the results and effects involving ultrasound are comparable. In addition, the authors systematically compared their results to those under silent conditions during several hours in the IL medium and under sonication in molecular solvents such as acetonitrile, methanol, ethanol, tetrahydrafuran (THF), dimethyl sulfoxide (DMSO), hexadecane, PEG-400 and/or dichloromethane.

The same synergetic effects were widely reported by other research groups for many organic reactions such as Heck arylation,[181] Mizoroki–Heck reactions,[182] Suzuki aryl–aryl cross-coupling,[183] Knoevenagel condensation,[184] Morita–Baylis–Hillman reaction,[185,186] Sonogashira coupling reactions,[187] the synthesis of aryl azides,[188] the synthesis of 2,3-disubstituted benzo[*b*]furans,[189] Michael addition of 1,3-dicarbonyl compounds to nitroalkenes,[190] Kabachnik–Fields reactions[191] and many others. Here, our goal is not to comment on all of the examples reported in the literature, but to identify the most important common features and mechanistic understanding. In all cases, similar advantages are described: (i) significant reduction of reaction time compared to silent conditions, (ii) improvement in yields and selectivities, (iii) need for lower catalyst loading, (iv) decrease in the volume of solvent medium needed and (v) ability to recycle the IL several times. As such, these ILs/ultrasound based processes offer significant advantages in terms of green chemistry.[192,193]

We especially worked on the epoxidation of cyclohexene, cyclooctene, styrene and α-pinene in the presence of a manganese tetraphenylporphine and hydrogen peroxide using the hydrophobic methyloctylpyrrolidinium bis(trifluoromethylsulfonyl)imide ([C_8C_1pyr][NTf$_2$]) under 20 kHz ultrasound irradiation (1 hour of irradiation).[194] Interestingly, we demonstrated a switch of reaction mechanism according to the chosen experimental conditions: (a) acetonitrile/silent stirring or (b) IL/ultrasound activation (Figure 15). In the first case, the metalloporphyrin was quickly degraded and its recycling was not possible. In the optimized conditions (b), the IL prevented the degradation of the catalyst. Moreover, the reaction time was significantly reduced under ultrasound.[195] In the latter conditions, the

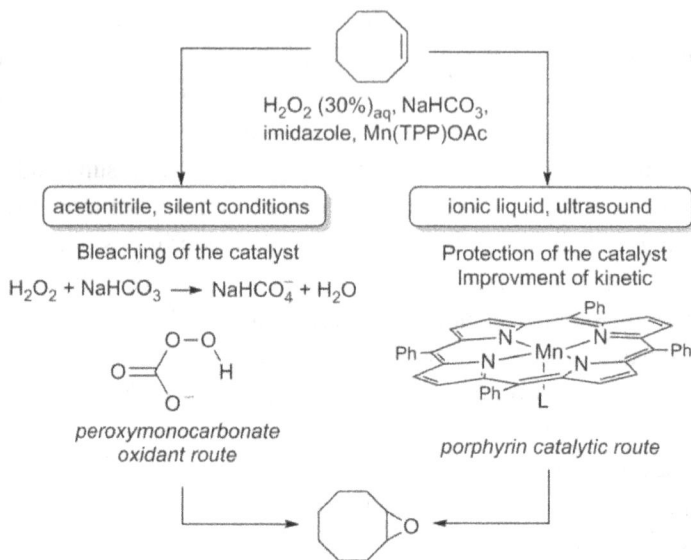

Figure 15. Proposed mechanisms for olefins epoxidation reaction (a) in acetonitrile/silent conditions and (b) in IL/ultrasound conditions.[195]

epoxidation reaction could occur *via* a classical high-valent oxo-manganese porphyrin complex. To unambiguously prove these mechanisms, a chiral Mn porphyrin complex was used as catalyst: in the classical conditions (a), the racemic mixture of epoxides was obtained while in the conditions (b), the ultrasonic asymmetric catalysis in IL clearly showed that the mechanism involved the metalloporphyrin catalyst route.[195] Thus, the IL/ultrasound combination not only improved the yields and decreased reaction times, but also involved a new reactivity for this epoxidation reaction.

In summary, it is clear that in a wide range of organic reactions, the IL/ultrasound combination is often described in catalytic reactions as improving the yield, the selectivity and the global efficiency of the process. However, the origins (mechanical versus chemical) of these outcomes may be several-fold depending on the irradiation power and the nature of the reaction.[192–202] Thus, it is important to not over-generalize on the impact of the IL/ultrasound combination. A valid comparison with classical conditions is required to prove the true impact of the IL/ultrasound combination. In addition, we recommend all researchers studying these fascinating effects to provide a clear set of control experiments to provide a basis for unambiguous conclusions.

6.1.2. Green synthesis of ionic liquids under ultrasound

In 2002, Varma's and Lévêque's groups proposed around the same time the use of ultrasound to improve the synthesis of ILs.[203,204] The first group developed a solvent-free sonochemical protocol for the preparation of some 1-alkyl-3-methylimidazolium halides ILs, comparing both the use of an ultrasonic bath (40 kHz, 320–881 W), with a probe system (20 kHz, 750 W) and oil bath conditions (Scheme 12).[203] From chloro-, bromo- and iodoalkanes and methylimidazole, excellent yields (mainly up to 92%) were obtained in substantially reduced times (from 0.25–6 hours to 25–34 hours in silent conditions).

Lévêque *et al.* reported the ultrasound-assisted preparation of several 1-butyl-3-methylimidazolium salts (BF_4^-, PF_6^-, $CF_3SO_3^-$ and BPh_4^-; Scheme 13).[204] While the maximal isolated yield (80–90%) was reached after 30 hours in silent conditions at room temperature, only 1 hour is necessary to reach these results under ultrasound (20–24°C maintained by a cooling bath, 30 kHz). The authors concluded that ultrasonic

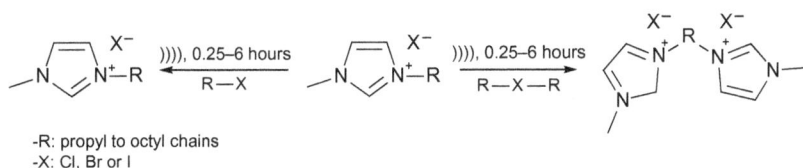

-R: propyl to octyl chains
-X: Cl, Br or I

Scheme 12: Ultrasound-assisted preparation of 1-alkyl-3-methylimidazolium halides (20 kHz, ultrasonic bath and probe system).[203]

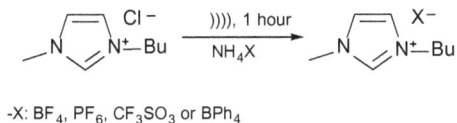

-X: BF_4, PF_6, CF_3SO_3 or BPh_4

Scheme 13: Ultrasound-assisted preparation of 1-alkyl-3-methylimidazolium salts (30 kHz, probe system).[204]

irradiation enhanced the anion exchange in the preparation of second generation ILs (the metathesis step).

A number of other reports including those of Varma,[205] Lévêque,[206,207] Zhao,[208] Cravotto,[206,207] Li[209] and more recently Messali[210,211] and Moldoveanu,[212] described sonication during IL syntheses that mainly leads to an important reduction of reaction time compared to classical methods. Some reviews have also discussed the use of low-frequency ultrasound or/and microwaves as activation methods for ILs syntheses.[213–215] Interestingly, Deetlefs and Seddon assessed the "green credentials" of the syntheses of ILs promoted by ultrasonic irradiation.[216] Indeed, the reduced preparation times induced by hotspots and small-sized bubble formation represent a significant green advantage compared to traditional methods, especially when the preparation is performed solvent-free. However, the authors noted that, as discussed above, the coloration and slight decomposition sometimes observed when ILs are exposed to ultrasound is an issue.[217,218] From an industrial point of view, this could be very limiting since the purification and decolorization of the salts that is required as a consequence leads to poor *E*-factors. Thus, the main challenge for the ultrasound-assisted synthesis of ILs could be scaling-up, taking into account both the issues of relating to sonochemistry and the slight degradation of the ILs under ultrasound.

6.1.3. Applications in material synthesis

In the material synthesis area, the combined use of ILs and ultrasound was investigated in the synthesis of methanofullerene derivatives (Scheme 14).[219,220] Indeed, unique chemical and physical properties of fullerenes can lead to important applications in medicine, optics and material science. Performed in $[C_4C_1im][BF_4]$ instead of THF as solvent, the yield and reaction activity were improved *via* an increase of the dehalogenation reaction rate. However, the authors did not explore the role of the IL/ultrasound combination in the process. It is important to note that the 2–3 days under ultrasonic irradiations is not efficient in terms of energy consumption, compared to magnetic stirring.

The IL/ultrasound combination has also been used for graphene sheet preparation. For example, Dai *et al.* prepared a high concentration, stable

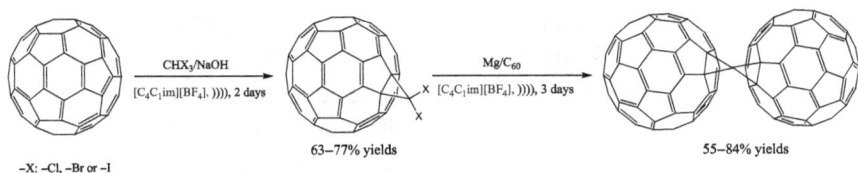

63–77% yields

55–84% yields

–X: –Cl, –Br or –I

Scheme 14: Ultrasound-assisted preparation of fullerenes in $[C_4C_1im][BF_4]$.[220]

graphene suspension in $[C_4C_1im][NTf_2]$ (up to 0.95 mg·mL^{-1}) from a dispersion of graphite, under 20 kHz ultrasonic irradiation (60 min).[221] Shi and Zhu reported an IL–Pd–graphene nanocomposite prepared *via* a sonoeletrochemical route, as an efficient electrochemical sensor for chlorophenols.[222] In this case, the ILs played the role of a linker[223] and enhanced the catalytic activity. Sonication presents known effects on the synthesis and modification of graphene, including the exfoliation of the graphite into discrete graphene sheets, suppressed aggregation in the reduction of graphite oxide compared to classical mechanical stirring and the promotion of the crystallization of nanoparticles by ultrasonic cavitation.[224] Liu *et al.* also reported the exfoliation of graphite into graphene sheets in 1-butyl-3-methylimidazolium cholate, and their stable dispersions that were achieved under 20 kHz ultrasonic irradiation.[225] They also applied this technique to achieve a Pd–nanoparticle–graphene hydrid, which was used as a catalyst for CO oxidation.[226]

Several preparations of nanocrystals such as ZnS, ZnO, Sb_2S_3 or SnS were performed in imidazolium-based ILs under 20–40 kHz ultrasound irradiation. The as-prepared nanocrystals were stabilized in the medium and ultrasonic treatment and allowed a decrease of preparation time compared to classical treatments.

Numerous papers have reported nanoparticle synthesis and stabilization in ILs,[227–229] and their sonochemical activation was observed in catalytic processes *via* dispersion improvement and surface depassivation.[230–233] Associating both technologies led to synergetic effects in many cases. For example, Jin *et al.* developed a sonochemical method for the preparation of gold nanoparticles capped by a thiol-functionalized IL using hydrogen peroxide as a reducing agent.[234] Here again, ultrasound (40 kHz, 80 W) accelerated the formation of gold nanoparticles and helped their dispersion in the IL. The function of the thiol groups in the selected IL was to prevent

Au^0 particles from aggregating and the 1-(2′,3′-dimercaptoacetoxypropyl)-3-methylimidazolium 3″-mercapto-1″-propanesulfonic acid IL controlled the subsequent growth of nanoparticles in the aqueous media, thanks to the thiol groups both in the cation and anion.[235] Behboudnia *et al.* applied their sonochemical preparation method in a 1-ethyl-3-methylimidazolium ethyl sulfate/water mixture for the synthesis of several nanoparticles such as SnO_2, CuS, PbS, CdS.[236–239] In all cases, the preparation was fast, efficient and led to very small and highly dispersed nanoparticles. Others examples have reported the same advantages in different ILs.[240–243]

6.1.4. Extraction and microextraction

For the last few years, extraction has been the first application of IL/ultrasound combination.[157] Here, we do not present all of the examples from the literature, since the technology is very similar in all cases, but instead try to probe the reasons for the synergistic properties in this context. In fact, ILs and ultrasound are widely used, separate extraction processes. Sometimes, heat-reflux extraction is laborious, time consuming and requires large amounts of volatile and hazardous organic solvents and ultrasound provides efficient solutions (see Section 5 of this chapter, page 95). On the other hand, ILs have also been investigated for extraction processes for their solvation properties, high chemical stability and the tuning opportunities that they offer.[244–246] Recently, ultrasonic-assisted extraction in ILs has become an efficient approach, especially to reduce the reaction time and facilitate procedures. For example, Cao *et al.* extracted piperine from white pepper *via* an ultrasonic pretreatment (bath, 40 kHz) in imidazolium-based ILs.[247] The procedure only consisted of treating the sample powder in a water/IL mixture with low frequency ultrasound, and after filtration and dilution, the solution was analyzed by UPLC. No effects attributable to the ILs were observed on peak resolution, elution order and elution time. Liquid-phase microextraction for determination of aromatic amines in water samples also used the same kind of procedure showing the performance, simplicity, stability, low cost, ease of operation and low consumption of organic solvents offered by this method.[248]

Thus, a synergistic IL/ultrasound combination has been extensively demonstrated in the efficient extraction and microextraction of organic compounds from liquid or solid products, coupled to different analysis techniques such as chemiluminescence detection,[249] GC-MS,[250] HPLC,[251–253] high-speed counter-current chromatography,[254] flame atomic absorption spectrometry,[255,256] liquid chromatography-quadrupole-linear ion trap-mass spectrometry,[257] fluorescence detection,[258] etc. The origins of this synergy are certainly related to mass transport effects at the micro and nanolevels, the ultrasound compensating for the elevated viscosity of the ILs.

6.1.5. Effect of ultrasound on non-volatile ionic liquids — Theoretical point of view

These numerous applications lead us to question more fundamentally about the process occurring in ILs under ultrasonic irradiation. What are the effects of ultrasound on non-volatile ILs? In 2003, Suslick's group investigated the sonochemistry and sonoluminescence of some imidazolium-based ILs [$(C_4C_1im)Cl$, $(C_4C_1im)(BF_4)$, $(C_4C_1im)(PF_6)$] and decyl-methylimidazolium tetraphenylborate [$(C_{10}C_1im)(BPh_4)$] under 20 kHz irradiation (60 $W \cdot cm^{-2}$) for 3 hours at 85°C and 135°C under an Ar flow.[217] During sonication, all the studied ILs darkened from colorless to amber indicating some decomposition. The IR spectra, ^{13}C NMR spectra, ^{19}F spectra, fast atom bombardment mass spectra (FAB-MS), UV–Visible spectra and elemental analysis of ILs contained no significant difference before and after sonication. However, the ^{1}H NMR spectrum obtained after sonication contained some additional peaks in the imidazole region amounting to 0.44% of total hydrogen. Interestingly, the headgas over each sonication was analyzed by GC-MS (Table 5). During sonication, the imidazolium-based ILs produced gases containing traces of light hydrocarbons and nitriles, clearly due to the degradation of the imidazolium rings. Headgases from sonication of [C_4C_1im][BF_4] and [C_4C_1im][PF_6] contained no detectable fluoride-containing species and from [$C_{10}C_1im$] [BPh_4] contained 72% benzene and traces of other cyclic products.

Suslick *et al.* also compared the multibubble sonoluminescence (MBSL) spectra of [C_4C_1im]Cl, 1-methylimidazole and 1-methylimidazole with 1.5% *n*-butyl chloride, showing molecular emission from excited states of C_2 carbon and CH (Figure 16). Like sonochemistry,

Table 5: Headgas composition during sonication of ILs.[217]

Entry	IL	Headgas components
1	$[C_4C_1im]$	chlorobutane (25.6%), chloromethane (51.1%), imidazole decomposition products (23.3%)[a]
2	$[C_4C_1im][BF_4]$	imidazole decomposition products[a]
3	$[C_4C_1im][PF_6]$	imidazole decomposition products[a]
4	$[C_4C_1im][BPh_4]$	benzene (71.6%), toluene (7.8%), cyclopentadiene (1.4%), 1-hexene (0.5%), 2,4-hexadiene (0.7%), imidazole decomposition products (18%)[a]

[a] Imidazole decomposition products: 1,3-butadiene (0.4%), 1,3-butadiyne (2.2%), acetonitrile/isocyanomethane (21.9%), 2-methylpropane (60.7%), 2-propenenitrile (7.4%) and pent-3-en-1-yne (7.4%).

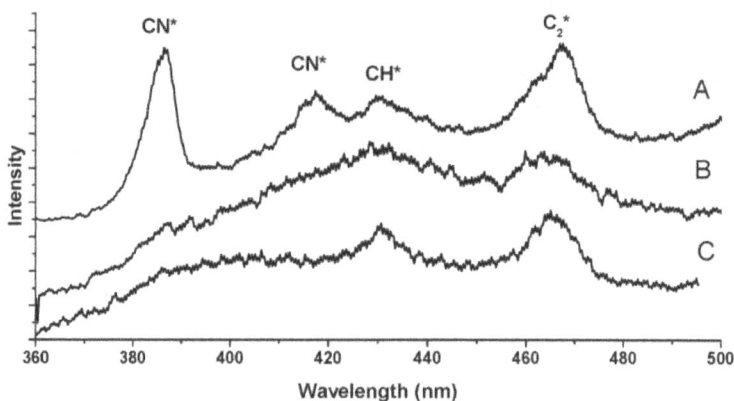

Figure 16: MBSL spectra of (A) 1-methylimidazole, (B) 1-methylimidazole with 1.5% *n*-butyl chloride and (C) $[C_4C_1im]$Cl. Reprinted with permission from Ref. 217. Copyright (2016) American Chemical Society.

sonoluminescence derives from acoustic cavitation, the implosive collapse of the bubbles, generating huge pulses of energy, which leads to the emission of photons.[259,260] They concluded from the products analyzed by ^{1}H NMR that the headgases and the MBSL spectra are a result of the ultrasonic decomposition of both the ILs themselves and of their primary sonolysis products.[217,261] The primary decomposition products for the imidazolium-based ILs are *N*-alkylimidazoles and 1-alkylhalides.

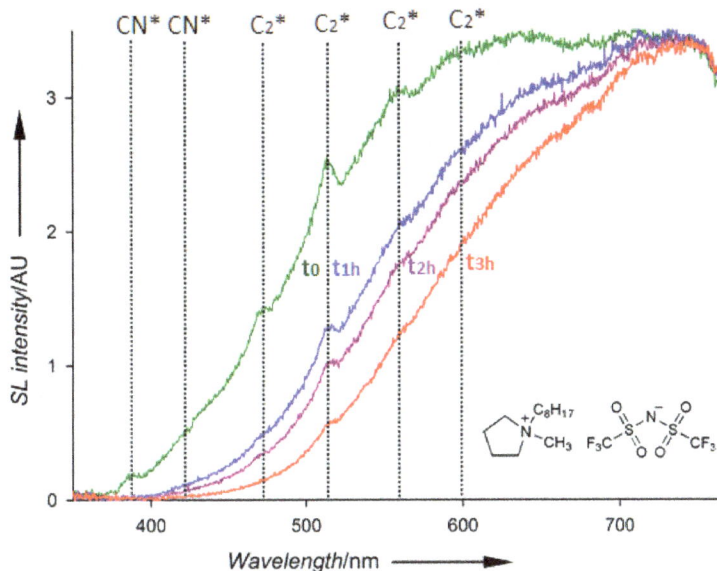

Figure 17: MBSL spectra of [C$_8$C$_1$pyr][NTf$_2$]. Reprinted with permission from Ref. 218. Copyright (2016) American Chemical Society.

In another study, we showed that hydrophobic bis(trifluoro-methylsulfonyl)imide (NTf$_2$)-based ILs decomposed under 20 kHz ultrasonic irradiation and identified the corresponding degradation products.[218] In fact, even if the proportion of these degradation products were limited (~ a few ppm), it was sufficient to prevent the recycling and the reuse of ILs in an organic reaction. As a theoretical point, the sonoluminescence spectra confirmed that cavitation occurs in pyrrolidinium- (Figure 17) and piperidinium-based ILs, showing the usual sonoluminescence continuum and molecular emissions from excited states of CN and C$_2$.

The mechanisms of sonochemical degradation of [C$_4$C$_1$im][NTf$_2$] were determined through the identification of the primary and secondary sonolysis products.[218] Pyrolysis reactions were suggested at the site of collapse of the cavitation bubbles. No product of oxidation by HO$^\bullet$ radicals was detected. Thus, analyses reported degradation products of the octyl chain (44%), benzene derivatives obtained through reforming mechanisms (20%), derivatives of acrylonitrile obtained from ammoxidation of propene under air (6%) and sulfur-containing compounds resulting from the degradation of the NTf$_2^-$ anion (3%). The temperatures and pressures required for pyrolysis reactions fit with the intense local heating (about 5,000 K) established when cavitation bubbles collapse.

The results obtained in this study[218] and Suslick's works[217,261] are consistent with the two-site model of sonochemical reactions involving the bubble's gas-phase interior and the immediately surrounding shell of liquid phase (see Chapter 2, page 13).[262] Ashokkumar *et al.* determined a temperature of about 3,500 K generated in the imploding cavitation bubbles in 1-ethyl-3-methylimidazolium ethylsulfate ([C$_2$C$_1$im][EtSO$_4$]) and observed an enhancement in the sonoluminescence intensity with increase in bulk fluid temperature and the corresponding decrease of the IL viscosity.[263]

A water/IL biphasic system was proposed to limit the degradation of hydrophobic ILs under ultrasonic irradiation, reducing by 20 times the amounts of degradation products.[218] It was explained by the hot spots that could occur preferentially in the aqueous phase rather in the IL, mainly due to the difference of viscosities and vapor pressures between these two media.

In another study, we determined for the first time, the acoustic power when some NTf$_2^-$-based ILs are submitted to ultrasound.[264] Despite very different specific heat capacities (c_p) for water and for ILs, the acoustic powers measured for the same electric power were very similar for both media. Thus, the faster heating up of the ILs (due to lower heat capacity) compared with water can lead to interesting effects as a solvent for organic reactions (Figure 18).

These studies highlight an important issue — that IL degradation may be a significant process in the application of the IL/ultrasound combination to organic reactions and cannot be ignored. Potentially, the IL will only be recyclable under carefully controlled conditions and these need to be identified by thorough identification of any breakdown products that appear. It is also important to recognize the possible role of these breakdown products in the enhancement of the chemical reactions taking place in the process.

Further investigations are needed to probe whether the cavitation model can be applied more broadly to all non-volatile ILs. On the other hand, distinctly volatile, protic ILs, in which proton transfer to form the

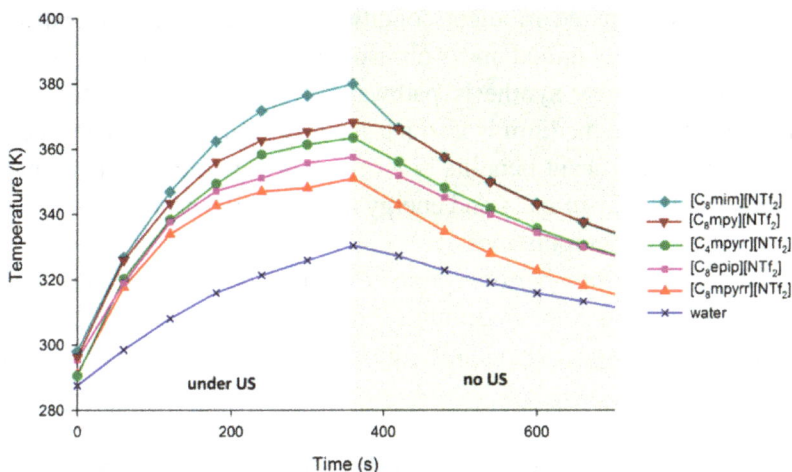

Figure 18: Temperature rise measured in some ILs and in water under ultrasound (P_{elec} = 11.5 W). Reprinted with permission from Ref. 264. Copyright (2016) American Chemical Society.

neutral acid and base compounds is possible, represent a completely different case in this context, and little is known about the effect of ultrasound on these ILs.

6.1.6. Conclusions about IL/ultrasound combination

In the two last decades, the use of ultrasound has increasingly been investigated in the presence of, or for the synthesis of ILs. The main areas where this synergistic combination are applied are organic chemistry, the synthesis of ILs and more recently for materials preparation (catalysts, nanoparticles, nanotubes, propellants, etc.), for extraction and microextraction, and for other applications (electrochemistry, sonochemical degradation of ILs, etc.). For instance, the use of low frequencies is predominant with IL media, probably for its ease of processing and its accessibility (ultrasonic baths or in some cases, ultrasonic probes). Indeed, the IL improves the physical effects of sonochemistry such as the generation of shockwaves, microjets, microconvection, microemulsions, erosion, etc. Most often, the beneficial consequences are the decrease of reaction/preparation times with improvements in yield, selectivity and/or quality of the products, compared to silent conditions. In some cases, unexpected results can also be obtained under ultrasonic irradiations, thereby offering the potential for new synthesis pathways. Increasingly, low frequency ultrasound in an IL medium is also particularly valuable in the extraction area, offering significant benefits in terms of processing time and energy reduction, efficiency in mass and energy transfer, high reproducibility and simplification of manipulations.

The increasing number of reported examples demonstrates the potential for further promising opportunities to emerge from the ILs/ultrasound combination. We especially identify the following important issues, trends and potentials:

(1) **Need of rigorous experimental data about ultrasound and ILs:** There is huge scope for more detailed investigation of the mechanisms involved in ultrasonic irradiation in ILs. As we have already indicated in the previous chapter of this book, it is true for all uses of ultrasound, but even more so when ILs are chosen as propagation media since the low volatility of the IL represents an extreme regime for the cavitation phenomena. The purity of ILs, the identification of impurities and the water content are very significant issues and should be clearly specified in all experimental sections.

(2) **Ultrasonic energy in the presence of ILs:** The most important challenge of the IL/ultrasound combination is to find the right balance in terms of acoustic power delivered to the reaction medium. At low frequencies, it is necessary to provide enough energy to favor the physical effects. In many cases, the highly heterogeneous and viscous systems obtained in ILs require a direct irradiative mode *via* an ultrasonic probe directly immersed in the solution, since ultrasonic baths are not always sufficiently powerful. However, depending on the IL nature and purity, an acoustic power that is too high can cause partial degradation of the IL (even at trace level). Many papers reported the darkening of the irradiated IL as a function of the time; this represents a real issue for industrial applications and might be sufficient to deem the process impractical.

(3) **Water amount in ILs under ultrasound:** The presence of water or organic impurities in irradiated ILs may impact significantly the acoustic cavitation phenomenon. As we explained previously, in a water/hydrophobic IL medium submitted to 20 kHz ultrasound irradiation, the hot spots occurred preferentially in water rather than in the IL due to the difference of viscosities and vapor pressures between water and IL.[264] This direction needs to be investigated in depth and could represent an opportunity to avoid the darkening and degradation of ILs under ultrasonic irradiation in many reported examples. Intentional spiking with water at low levels is recommended as a vital supplementary experiment in all reports that involve ILs, in order to demonstrate the positive and/or negative effects of the presence of water. The choice of the IL nature and purity could also be an essential parameter to impact the cavitation and limit the degradation.

(4) **Investigation of high frequency ultrasound:** High frequency ultrasound seems currently to be mostly of interest as a means for IL degradation *via* advanced oxidation processes. For example, Li *et al.* developed an efficient process for oxidative degradation of 1,3-dialkylimidazolium ILs in hydrogen peroxide and acetic acid medium under high frequency ultrasound (330 kHz, 750 W).[265] The authors achieved a degradation efficiency of 93% after 12 hours, and 99% after 72 hours. Degradation products were determined using GC-MS to propose a possible mechanism of degradation (Scheme 15). In fact, three hydrogen atoms in the imidazolium ring were preferentially oxidized, followed by cleavage of the alkyl

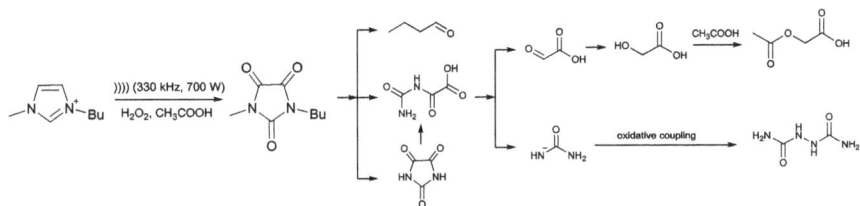

Scheme 15: Oxidation degradation mechanism of $C_4C_1im^+$ cation under high frequency ultrasound.[265]

groups attached to the *N* atoms from the ring. Selective fragmentations of C–N bonds of the imidazolium ring lead to ring opening, forming degraded intermediates. Thus, it appears that the US irradiation provides a valuable approach to the destruction of ILs in, at least small scale, waste streams and has the potential to be scaled up to a continuous larger scale process by appropriate engineering design.

The use of high frequency ultrasound needs to be further explored in order to provide a basis for optimizing reactivity in the presence of IL-based media for organic and catalytic reactions.

(5) **New opportunities for IL/ultrasound combination:** This combination opens the door to interesting opportunities in new research areas. For example, this combined technology produces exciting results for biomass pretreatment.[266–273] The combination with other innovative technologies represents a great potential in terms of innovation, for example, coupling with enzymatic catalysis for enhancing enzymatic activity in ILs.[274,275] In these cases, it will be important to study silent, coupled and non-coupled conditions separately for comparison and to highlight the synergetic effects.

Low frequency ultrasound activation appears also to improve electrochemical processes mainly *via* stirring effects (increase of mass-transfer coefficients),[276] continuous cleaning of the electrode surface and enhancement of reaction rates.[277] For these reasons, some examples of ultrasonic-electrodeposition processes in ILs were reported by Zeng's[278–281] and Zheng's[282,283] research groups.

Based on the numerous advantages discussed here, in terms of time reduction, yield improvement, energy economy and innovation, this US/IL combination clearly has a strong potential to contribute to innovation broadly in green and sustainable chemistry.

6.2. Ultrasound coupled to microwave irradiation

Electromagnetic radiation with a frequency in the range 0.3–300 GHz (microwave, noted MW) heats matter through a dielectric mechanism that may involve dipolar polarization and ionic conduction. It is the ability of a material to absorb microwave energy and convert it into heat that causes bulk heating; the temperature of the whole sample rises simultaneously, contrary to conventional conductive heating.[284,285] Although non-thermal effects have been invoked for reactions in which the transition state is more polar than the ground state, it is generally accepted that rate enhancements largely stem from thermal effects.[286,287]

The superior heating provided by microwave associated to attractive physical effects brought by ultrasound represents an innovative way in synthesis and processing.[288] Based on the recent literature, Cravotto and Cintas reported the synergetic effects provided by these two technologies, even if the mechanisms of cavitation and microwave effects are not fully understood.[289]

As described in Chapter 4 (page 41), the ultrasonic energy generated by a transducer is delivered to the reaction vessel by means of a horn, usually made of titanium alloy. Evidently, a piece of metal placed inside a microwave chamber will lead to arcing that could result in vessel rupture or perhaps an explosion, if flammable compounds are involved (Please do not test this in your domestic microwave oven!). The conditions of temperature, pressure, stirring and power input have to be controlled during microwave irradiation to avoid the formation of electric arcs.[290,291] To overcome these limitations, ultrasonic irradiation can be achieved inside a modified MW oven by inserting in it a horn made of quartz, even though quartz is not the ideal piezoelectric material because of its fragility. Pyrex also shares the same drawback, while ceramic horns are more expensive. The most recent development introduces horns made from engineered plastics, such as polyether ether ketone (PEEK) or polytetrafluoroethylene (PTFE), materials that are much more resistant to shock and can be more firmly jointed to the booster (Figure 19).[292]

Some problems of reproducibility occurred using modified domestic ovens, and the control of reaction parameters is not quite accurate (Figure 20). For these reasons, some professional multimode systems available on the market were designed (Figure 21).

Combined ultrasound and microwave irradiations can be performed **simultaneously** or otherwise **sequentially** by circulating the reacting mixture in two loop reactors. Indeed, loop reactors and flow systems are particularly attractive for selective activations (Figure 22) and show promise as automation may cause batch microwave- or ultrasound-based chemistries to develop into cleaner and more efficient continuous processes.[293]

Figure 19: Different types of microwave-inert ultrasonic probe made in quartz, pyre and PEEK. Reprinted with permission from Ref. 289. Copyright (2016) John Wiley & Sons, Inc.

Figure 20: A modified domestic oven for simultaneous ultrasound and microwave irradiation with a quartz horn. Reprinted with permission from Ref. 289. Copyright (2016) John Wiley & Sons, Inc.

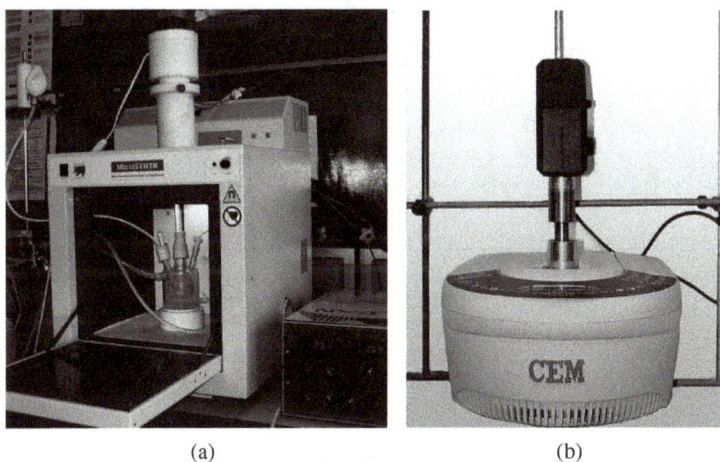

(a) (b)

Figure 21: (a) Simultaneous irradiation in a multimode microwave oven (Milestone®) equipped with a pyrex horn. Reprinted with permission from Ref. 289. Copyright (2016) John Wiley & Sons, Inc. (b) Hybrid reactor with a Pyrex horn inside of a microwave zone formed *via* a circular wave guide. Reprinted with permission from Ref. 288. Copyright (2016) Elsevier.

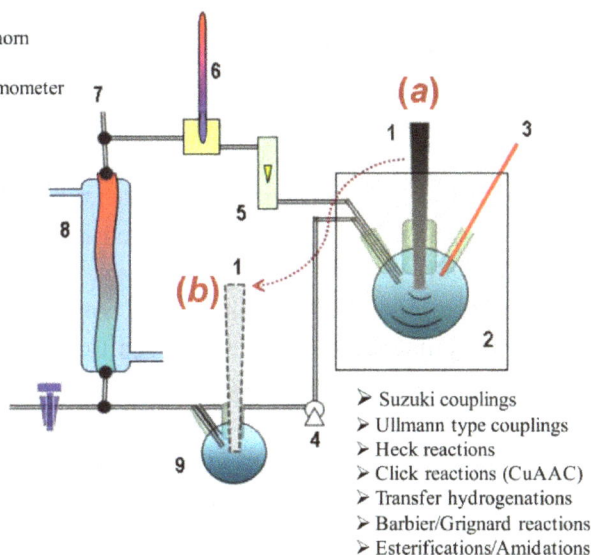

1. US non-metallic horn
2. MW oven
3. Optical fibre thermometer
4. Pump
5. Flow-meter
6. Thermometer
7. Inlet and sampler
8. Heat exchanger
9. External flask

➤ Suzuki couplings
➤ Ullmann type couplings
➤ Heck reactions
➤ Click reactions (CuAAC)
➤ Transfer hydrogenations
➤ Barbier/Grignard reactions
➤ Esterifications/Amidations

Figure 22: Hybrid loop reactors with two different configurations: (a) simultaneous ultrasound/MW irradiation; (b) sequential ultrasound/MW irradiation in separate vessels. Reprinted with permission from Ref. 293. Copyright (2016) Elsevier.

Table 6: Hydrazinolysis of methyl salicylate.[303]

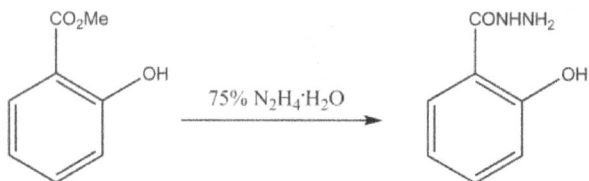

Method	Reaction time	Yield (%)
Reflux	9 hours	73
US (20 kHz, 50 W) + reflux	1.5 hours	79
MW (2.45 GHz, 200 W)	18 min	80
MW + US	40 s	84

Based on the numerous advantages provided by this combination, recent applications were reported in chemistry, in particular, for the synthesis of catalysts or materials,[294–296] to perform catalytic or organic reactions[297–299] and for extraction processes.[300–302] We will not describe all the examples of the literature here, but we will illustrate the advantages of the coupled technology, through few examples.

The first example is the hydrazinolysis of esters with hydrazine monohydrate under solventless conditions using a modified domestic oven.[303] Table 6 shows the dramatic reduction of reaction times (84% yield within 40 s). As these transformations occurred under heterogeneous conditions, acceleration was interpreted in terms of enhanced heat and mass transfer.

The application of the MW/ultrasound combination for extraction is also of primary interest. In this area, high throughputs, low consumption of reagents, total automation and increased safety represent essential goals. Although both microwave and ultrasound may cause undesired chemical transformations or degradation, the combination of these two energy sources can promote and improve extraction process of natural products.[42]

Hernoux-Villière *et al.* studied the catalytic conversion of starch-based industrial waste into sugars using a simultaneous microwave-assisted procedure.[304] Two hours of combined irradiation in sulfuric acid led to 46% conversion into reducing sugars. In the same conditions, ultrasound led to 1% yield and microwave to 35% yield. The synergetic effect of the combination was explained by the heat transfer of energy in a significant short time period provided by microwave and the efficient mass transfer of low frequency ultrasound irradiation (20 kHz).[305]

6.3. Sono-assisted electrochemistry

Historically, the use of ultrasound in combination with electrode processes and corresponding applications, called *sonoelectrochemistry*, started in the 1930s[306] and was widely reviewed in the literature.[307–311] The main topics in the field of sonoelectrochemistry have been identified by Pollet in his book[312]:

⇨ use of electrochemistry as a tool to investigate cavitation bubble dynamics,
⇨ sonoelectroanalysis,
⇨ sonoelectrochemistry in environmental applications,
⇨ organic sonoelectrosynthesis,
⇨ sonoelectrodeposition,
⇨ influence of ultrasound on corrosion kinetics and its application to corrosion tests,
⇨ sonoelectropolymerization,
⇨ sonoelectrochemical production of nanomaterials,
⇨ sonoelectrochemistry in hydrogen and fuel cell technologies.

In this subsection, we will not report the complete review of the literature, but rather some interesting points in the field, showing the interest of this combination.

Hu *et al.* reported an interesting sonoelectrochemical preparation of gold nanoparticles and carbon nanosheets hybrid (Figure 23).[313] Indeed, the proposed method involved simultaneous generation of carbon nanosheets by oxidation of graphite anode and Au nanoparticles by reduction of $AuCl_4^-$ on the surface of cathode. Then, the Au nanoparticles modified with poly(diallyl dimethyl ammonium chloride) were self-assembled on the surface of carbon nanosheet hybrids. Here, the advantages of the

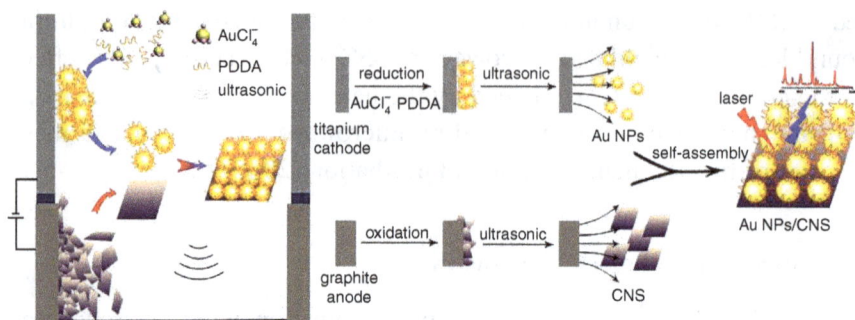

Figure 23: Schematic illustration of the fabrication of Au NPs/CNS by the one-step sonoelectrochemical method. Reprinted with permission from Ref. 313. Copyright (2016) Royal Society of Chemistry.

synergy effect of electric field and ultrasonic field were: (i) the possible control of size and distribution on the surface of carbon nanosheets, (ii) the high quality of the surface-enhanced Raman scattering (SERS) activity. Here, ultrasound irradiation improved the dispersion of the electrolytic solution. In addition, the mixing effect of ultrasound allowed the rapid transfer from the cathode and anode vicinity to the bulk solution.

Another interesting example is the sonoelectrodeposition of copper(II) chloride on Pt electrodes in aqueous and deep eutectic solvents (glycerine 200) using 20 kHz and 850 kHz as ultrasonic frequencies.[314] Authors showed that the deposition of copper was greatly affected by ultrasound. Limiting current densities were obtained in both solvents under sonication at 20 kHz and 850 kHz and a 10-fold and a 5-fold increase in currents in aqueous potassium chloride and glycerine 200 compared to silent conditions were observed, respectively.

The reduction of copper was also investigated in other deep eutectic solvents such as choline chloride-ethylene glycol.[315] Both cyclic voltammetry and linear voltammetry were performed at three temperatures (25°C, 50°C and 80°C) and under ultrasonic conditions (20 kHz, 5.8 W) to calculate the mass transfer diffusion coefficient (k_D) and the standard rate coefficient (k). These parameters are used to determine that copper reduction is carried out *via* a mixed kinetic-diffusion control process. Temperature and ultrasound showed the same effect on mass transfer for reduction of Cu^{II}/Cu^{I}. However, temperature was more beneficial than

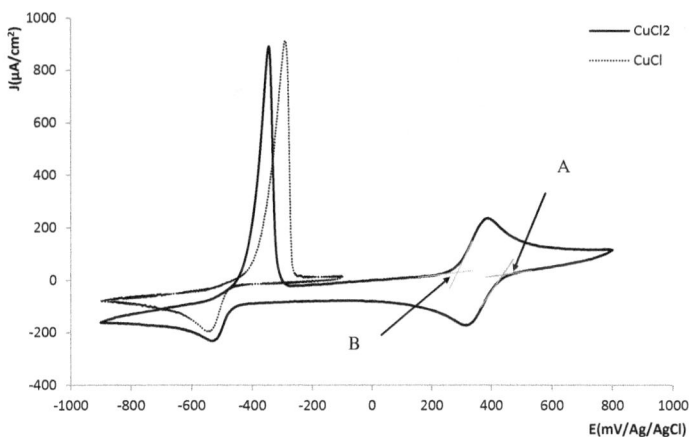

(a)

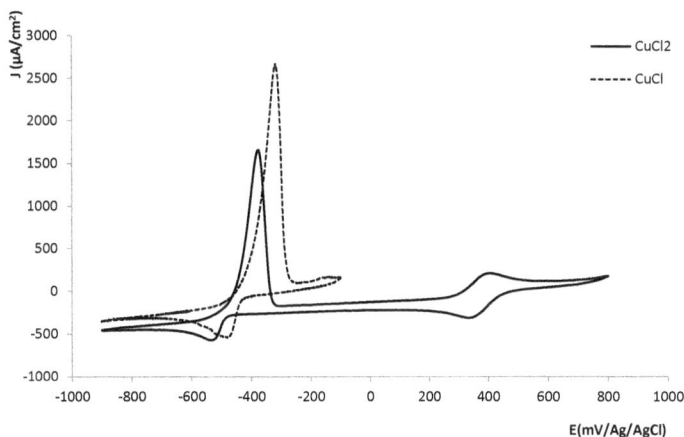

(b)

Figure 24: Voltammograms for CuCl (dotted line) and CuCl$_2$ (solid line) in choline chloride-ethylene glycol under 20 kHz ultrasound at a 10 mV/s scan rate: (a) 25°C (b) 80°C. Reprinted with permission from Ref. 315. Copyright (2016) Elsevier.

ultrasound for mass transfer of Cu^I/Cu^0. Standard rate constant improvement due to temperature increase was of the same order as that obtained with ultrasound. But, by combining higher temperature and ultrasound, reduction limiting current was increased by a factor of 10 compared to the initial conditions (25°C, silent conditions), because ultrasonic stirring is more efficient in the lower viscosity fluid (Figure 24). Interestingly, these

121

values can be considered as key parameters in the design of copper recovery in global processes using ultrasound.

Mason and Saez Bernal described the different parameters that may influence sonoelectrochemistry[316]:

(1) **Acoustic streaming:** The main effect of the acoustic streaming is to promote the movement of the solution, enhancing mass transfer of electroactive compounds to the electrode surface and to reduce the diffusion boundary layer.

(2) **Turbulent flow:** The movement of the cavitational bubbles under the influence of the ultrasound field can cause a chaotic and disorderly movement of the solution, giving rise to a turbulent flow. The levels of turbulence are high near to the emitting surface and decrease rapidly with increasing distance from it. Turbulent flow is effective in improving mass transport processes within the solution and at the electrode surface and thus it can cause similar effects to acoustic streaming.

(3) **Microjets:** If the collapse of cavitation bubble takes place close to the surface of the electrode, the created liquid jet may directly impinge upon it.[317] The combined effects of microjets and microstreaming contribute to the enhanced mass transport of electroactive compounds to the electrode. Other consequences of the formation of microjets are electrode cleaning and surface activation that prevent fouling of the electrode surface and can also improve any electrodeposition processes.

(4) **Shock waves:** Another mechanical effect arising from cavitation in addition to shear forces and microjets in the bulk liquid is a shock wave generated in the last stages of the violent collapse. The attenuation of this pressure wave is rapid and the effects on the electrode surface will only take place when the collapse occurs close to it.

(5) **Chemical effects:** In addition to the mechanical effects, sonochemical effects with the formation of reactive radical species from the sonolysis of the electrolyte have been observed in several electrochemical processes. However, the influence of chemical effects in sonoelectrochemistry has been relatively little studied in comparison with the physical effects referred to above.

6.4. Ultrasound and enzyme combination

In their review, Bremner *et al.* listed and discussed the ultrasound-assisted bioprocesses in the literature to understand the influence of ultrasound irradiation on discrete enzyme systems, enzymes used in bioremediation, microbial fermentations and enzymatic hydrolysis of biopolymers.[318] It is clear that the generation of cavitation bubbles causes damages to biological molecules both *in vivo* and *in vitro*. Indeed, if low-power ultrasound increases growth in microbial cell cultures, high power causes cell disruption and is considered as microbicidal. However, the most recent works are now focused on the use of *sub-lethal doses* of ultrasonic irradiation, leading to beneficial effects in many bioprocesses.

The mechanism of action remains to be elucidated and probably varies depending on the substrates involved and the frequency and power of the ultrasound inputted, however, several mechanisms were proposed.[318] The most common suggestions are that ultrasound enhances physical phenomena (enhancement of emulsification, decrease of particle size, decrease of the unstirred diffusion layer, better mixing and/or microstreaming and mechanical stress). Other authors suggested that ultrasound (i) promotes cell membrane permeability and alters surface potential resulting in activation of calcium channels; (ii) weakens the cell wall/membrane promoting permeability and selectivity and accelerating nutrition or molecule transport and/or (iii) increases the dissolution rate of moderately soluble compounds and enhances mass transfer inside and outside of the cell. For biphasic system reactions, it is suggested that ultrasound lessens the adsorption of organics on the enzyme surface to ensure easy processing and recycling of the immobilized enzyme.

Some authors claim that controlled cavitation is not always detrimental and suggest that cavitation effects enhance the transport of enzyme macromolecules to the surface of the substrate which results in the opening up of the substrate surface to the action of enzymes as a result of mechanical impact of cavitation.

The main identified advantages in the use of the ultrasound/enzyme conditions are[318]:

(1) Lower consumption of expensive enzymes;
(2) Shorter processing times;

(3) Secondary structure of the enzyme generally is not affected, although the microenvironments and tertiary structure of the enzyme may be perturbed;

(4) For filamentous fungi, ultrasound can alter morphology and broth rheology without affecting growth;

(5) In ultrasound-enhanced enzymatic synthesis in organic solvents, the addition of small amounts of water can greatly increase reaction rates.

To date, most of the published research has concentrated on the low frequency ultrasound, but research investigating higher frequencies is beginning to appear with some promising results. For example, Wang *et al.* reported the disruption of microalgal cell using 3.2 MHz ultrasonic irradiation (horn, 100 W, controlled temperature) for lipids recovery.[319] They showed that with high frequency, the algal cell disruption was as effective as with low frequency (20 kHz) but with a lower energy use. Another example reported the use of 850 kHz ultrasonic irradiation at 60 W, 90 W, and 120 W in the presence of laccase. These experiments showed that, up to an energy input of 90 W, no enzymatic deactivation was observed in comparison to the activity of the native laccase, while increasing the power input to 120 W increased deactivation, leading to short half-life times (5 hours instead of 20 hours).[320,321]

The use of combined enzyme/ultrasound bioprocessing has been widely investigated for chemical production from lignocellulosic biomass. Yachmenev *et al.* suggested the following conclusions for the field: (1) cavitation effects enhancing the transport of enzyme macromolecules to the surface of the substrate; (2) opening up of the substrate surface to the action of enzymes as a result of mechanical impact of cavitation; (3) enhancement of the cavitation phenomena in heterogeneous systems; and (4) optimum enzyme temperature of 50°C coincides with maximum cavitation.[322] For example, the enzymatic hydrolysis of corn stover to glucose was increased up to 22% after 1 hour of combined enzyme/ sonication bioprocessing in comparison to silent conditions.

Table 7 reports some examples of enzymatic ultrasound-assisted reactions for chemical production from biomass feedstocks.[115] As explained below, the chosen conditions are low frequency ultrasound irradiation (<50 kHz) to favor the physical effects of ultrasound on enzyme activation

Table 7: Enzymatic ultrasound-assisted reactions for chemical production from biomass feedstocks.

Entry	Biomass origin	Ultrasonic conditions	Experimental conditions	Enzyme	Main observations	Ref.
1	Powdered crystalline cellulose	38 kHz, 80 W (ultrasonic bath, pulsed and continuous)	5–120 min, 21–35°C, pH: 4.8–5.5	*T. viride* cellulase	Ultrasound increases the rate of glucose production	324
2	Paper pulps (2.6–30.9% lignin)	20 kHz, 0–45 W (probe) + mechanical stirring	48 hours, 45°C, pH:4.8	*T. viride* cellulase	Saccharification of the unbleached kraft pulp from coniferous tree (2.5% lignin) efficiently enhanced by the continuous ultrasonic irradiation	325, 326
3	Corn stover	20 kHz, 80 W (ultrasonic cell crusher)	90 min, 25°C, pH: 5.5 (pulse)	Cellulase	Saccharification rate of the ultrasonically assisted enzymatic catalysis hydrolysis was enhanced by 70% compared to the conventional enzymatic catalysis hydrolysis	327
4	Corn stover and sugar cane bagasse	50 kHz, 13 A (Hexagon reactor, no pulse mode)	8 hours, 35°C, pH: 5.0	Trichoderma reesei cellulase	Enzyme efficiency was significantly improved under ultrasound by increasing the glucose rate	322
5	Envasive weeds	35 kHz, 35 W (ultrasonic bath)	10 hours (10% duty cycle), 30°C, pH: 5.0	Trichoderma reesei cellulase	Acceleration of hydrolysis kinetics by 10-fold under ultrasound	323

125

and on the reaction medium. Enzyme stability requires specific experimental conditions, such as the control of temperature (generally between 25°C and 45°C) and the pH of the reaction medium (generally between 4.8 and 5.5). Globally, the ultrasound/enzyme combination increased chemical production or reduced reaction time (Table 7). However, comparison between each example is not accurate, since biomass feedstocks, frequencies, powers, enzymes and experimental conditions are different.

Aliyu and Hepher showed that an ultrasonic bath at 38 kHz led to an increase of glucose production (0.28 ppm · min⁻¹) from powdered crystalline cellulose in the presence of *T. viride* cellulose (Table 7, entry 1).[324] The equipment used (ultrasonic bath) and the lack of rigorous comparison with silent conditions make difficult the understanding of ultrasound effects in this case. Moreover, the whole range of degradation and molecular rearrangement products obtained in addition to glucose were not identified.

Li *et al.* proposed a rigorous study on enzymatic saccharification of various waste papers, comparing ultrasonic and silent conditions (Table 7, entry 2). Even if the irradiation time is relatively long (48 hours), they compared the system to a simplified kinetic model and suggested a mechanism to explain the almost complete conversion of the pulp to sugar under ultrasound. Indeed, the predominant steps would be the stagewise formation of the reaction zone from the outer surface of cellulose fiber and the enhanced adsorption and desorption of enzyme for cellulose polymer chain in the reaction zone of the fiber.[325,326]

The saccharification rate from corn stover was also enhanced by low frequency ultrasound (Table 7, entries 3 and 4) by 70% compared to the conventional enzymatic catalysis hydrolysis.[322,327] In addition, the reduction of both time and concentration required for the alkali pretreatment and the increase of the removal rate of lignin were reported. Interestingly, sonication during the enzyme processing did not reduce the specific activity of enzyme macromolecules in any significant way.[322]

Gogate *et al.* suggested that the enzyme/ultrasound combination can be an efficient tool for bioethanol production. Indeed, among the four essential steps of lignocellulosic biomass conversion into bioethanol (pretreatment, hydrolysis, fermentation and distillation), they showed that ultrasound can be used for the two first steps (i.e., pretreatment and hydrolysis).[328]

Based on the state of art of enzymatic hydrolysis under ultrasound, they discussed the sonochemistry and cavitation effects on enzymatic process intensification through enhancement of enzyme diffusion, decrease of barriers of transportations, stabilization of enzyme in solution, influence of enzymatic reactivity, improvement of mixing and mass transfer. Very recently, they determined the ideal ultrasonic parameters (intensity and time) to enhance cellulose digestion by enzyme at low frequency.[329] Interestingly, the efficiency of ultrasound/enzyme combination was shown for the yeast fermentation step, accelerating the glucose fermentation to ethanol.[330,331]

A discussion about the activation and deactivation effects of ultrasound on enzyme is reported in the book of Mawson *et al.* in 2011.[332] Indeed, all the environmental conditions required for enzyme stabilization (temperature, pressure, shear stress, pH, ionic strength) are important both in silent and ultrasonic conditions. Overall, the advantages of ultrasound-assisted enzymatic processes are essentially the time reduction and the increase of yields. However, further studies are required to understand the mechanisms under ultrasound, depending of the substrate, the enzyme, the experimental conditions and the sonochemical parameters.

7. Environmental Remediation

If the previous applications of this chapter were essentially focused on the selective aspects provided by sonochemical effects, the ultrasonic degradation of pollutants in liquid effluents until total mineralization has also been widely investigated in the environmental remediation area. Green chemistry principles are rather focused on the prevention of pollution, but we will give some words on environmental remediation which can be considered as a solution for degradation of products or pollutants. It is essential to develop clean processes for that. In this case, sonochemistry can be considered as an advanced oxidation process, particularly since ultrasound provides HO• radicals.[333,334]

Indeed, high-frequency ultrasound applied in aqueous solutions induces the formation and the collapse of cavitation bubbles of high energy. In these conditions, the compounds present in the gas phase of imploding bubbles (poorly hydrophilic and/or volatile solutes and water vapor) undergo pyrolysis processes.[335] Through sonolysis, HO• and H• are produced with high concentration in a confined volume. Degradation processes of solutes involve either pyrolysis of gas-phase molecules within the bubble, or migration of HO• to the solution bulk, where transformation of hydrophilic and/or poorly volatile compounds can take place.[336]

Interestingly, Adewuyi listed all the radical species formed and recombination reactions that occurred for sonolysis (ultrasound), sono-Fenton and sonophoto-Fenton systems (in the presence of iron ions), for the coupling between ultrasound, ozone, hydrogen peroxide and UV-system and wet oxidation systems.[337] He also reported the combination of ultrasound with the addition of some gases, ions, acids or heterogeneous catalysts for the treatment of pollutants in water.[338] In all these strategies, the goal is to produce active and efficient radical species to degrade the organic molecules present as traces levels.

The sono-Fenton effect was recently studied for the disintegration of wastewater treatment plant sludge.[339] The results demonstrated that the coupling between ultrasound (20 kHz) and Fenton treatment ($FeSO_4$ + H_2O_2) increases soluble chemical oxygen demand, total organic carbon and extracellular polymeric substances concentrations in sludge supernatant. This method was more effective than ultrasonic or Fenton oxidation treatment alone: (i) increase of the release of soluble chemical oxygen demand by 2.1- and 1.4-fold compared with ultrasound and Fenton system alone, respectively; (ii) increase of the release of extracellular polymeric substances by 1.2-fold compared with ultrasound alone; (iii) finer particle size and looser microstructure; (iv) the concentration of hydroxyl radicals (HO•) increased from 0.26 mM by Fenton treatment to 0.43 mM by combined treatment. In summary, the combined treatment improved the release of organic matter from sludge.

Ultrasound was investigated for the treatment of different pollutants such as chlorinated compounds, phenolic compounds (bisphenol A, 4-chlorophenol, etc.), carboxylic acids, polymers, dye wastewater, pharmaceutical compounds and pesticides.[334] What is important to retain for this field of research is that ultrasound is often combined with other oxidative methods, showing synergetic effects and great improvement in efficiency.

8. Scale-up and Industrial Applications

The progress of sonochemistry in Green Chemistry is dependent upon the possibility of scaling up the excellent laboratory results for industrial use.[340] Section 4.4 from the chapter entitled *Ultrasonic Equipment* (page 41) showed different examples of reactors and industrial applications (extraction, generation and deposition of nanoparticles, production of pharmaceutical nanoemulsion, etc.). Three systems could be envisaged:

- Immerse reactor in a tank of sonicated liquid — indirect irradiation;
- Immerse an ultrasonic source directly into the reaction medium — direct irradiation;
- Use reactor constructed with vibrating walls — direct irradiation.

The development of sonochemical application at larger scale is now a crucial challenge for sonochemists to reproduce the exciting results obtained at lab into continuous or large-scale processes.[341] Sonochemistry has suffered from problems of reproducibility for a long time, especially with ultrasonic baths having an odd geometry in which frequency and power depend on the transducer employed. Likewise, ultrasonic probes and standardized systems will become more and more the instruments of choice.[42] A series of commercially available ultrasonic reactors can be directly adapted for scale-up,[342,343] giant probe systems using a large bar of steel as horn found different applications in processing and catalysis[344] and tubular reactors have already provided interesting results in continuous flow.[341] A challenge is to follow closely to see the future development of green processes based on sonochemistry in industrial applications.

References

1. D. H. Bremner, *Ultrason. Sonochem.* **1994**, *1*, S119–S124.
2. T. Mason, *Chem. Soc. Rev.* **1997**, *26*, 443–451.
3. J.-L. Luche, *Synthetic Organic Sonochemistry*, Plenum Press, New York, 1998.
4. M. Draye, J. Estager, E. Naffrechoux, J.-M. Lévêque, Organic sonochemistry, In: *Ultrasound and Microwaves: Recent Advances in Organic Chemistry* (Eds.: J. P. Bazureau, M. Draye), Research Signpost, 2011, 1–34.

5. G. Cravotto, E. Borretto, M. Oliverio, A. Procopio, A. Penoni, *Catal. Commun.* **2015**, *63*, 2–9.
6. J.-L. Luche, J.-C. Damiano, *J. Am. Chem. Soc.* **1980**, *102*, 7926–7927.
7. T. Ando, S. Sumi, T. Kawate, J. Ichihara, T. Hanafusa, *J. Chem. Soc. Chem. Commun.* **1984**, *31*, 439–440.
8. J.-L. Luche, C. Einhorn, J. Einhorn, J. V. Sinisterra-Gago, *Tetrahedron Lett.* **1990**, *31*, 4125–4128.
9. T. J. Mason, *Ultrason. Sonochem.* **2003**, *10*, 175–179.
10. G. Cravotto, P. Cintas, *Chem. Sci.* **2012**, *3*, 295–307.
11. D. Chen, S. K. Sharma, A. Mudhoo, *Handbook on Applications of Ultrasound: Sonochemistry for Sustainability*, CRC Press, USA, 2011, 739.
12. S. Puri, B. Kaur, A. Parmar, H. Kumar, *Curr. Org. Chem.* **2013**, *17*, 1790–1828.
13. R. Patil, P. Bhoir, P. Deshpande, T. Wattamwar, M. Shirude, P. Chaskar, *Ultrason. Sonochem.* **2013**, *20*, 1327–1336.
14. M. Nikpassand, M. Mamaghani, F. Shirini, K. Tabatabaeian, *Ultrason. Sonochem.* **2010**, *17*, 301–305.
15. W. Liju, K. Ablajan, F. Jun, *Ultrason. Sonochem.* **2015**, *22*, 113–118.
16. R. Cella, H. A. Stefani, *Tetrahedron* **2009**, *65*, 2619–2641.
17. K. Wei, H.-T. Gao, W.-D. Z. Li, *J. Org. Chem.* **2004**, *69*, 5763–5765.
18. E. Pelit, Z. Turgut, *Ultrason. Sonochem.* **2014**, *21*, 1600–1607.
19. T. Javed, T. J. Mason, S. S. Phull, N. R. Baker, A. Robertson, *Ultrason. Sonochem.* **1995**, *2*, S3–S4.
20. M. Avalos, R. Babiano, N. Cabello, P. Cintas, M. B. Hursthouse, J. L. Jiménez, M. E. Light, J. C. Palacios, *J. Org. Chem.* **2003**, *68*, 7193–7203.
21. J. L. Bravo, I. López, P. Cintas, G. Silvero, M. J. Arévalo, *Ultrason. Sonochem.* **2006**, *13*, 408–414.
22. P. Nebois, Z. Bouaziz, H. Fillion, L. Moeini, J. A. Piquer, J.-L. Luche, A. Riera, A. Myanot, M. A. Pericàs, *Ultrason. Sonochem.* **1996**, *3*, 7–13.
23. S. H. Banitaba, J. Safari, S. D. Khalili, *Ultrason. Sonochem.* **2013**, *20*, 401–407.
24. F. Wang, G. Wang, W. Sun, T. Wang, X. Chen, *Microporous Mesoporous Mater.* **2015**, *217*, 203–209.
25. A. Khorshidi, *Chin. J. Catal.* **2016**, *37*, 153–158.
26. D. Azarifar, S.-M. Khatami, Z. Najminejad, *J. Iran. Chem. Soc.* **2014**, *11*, 587–592.
27. A. Akbari, M. Omidkhah, J. T. Darian, *Ultrason. Sonochem.* **2015**, *23*, 231–237.
28. M. Mirza-Aghayan, N. Ganjbakhsh, M. M. Tavana, R. Boukherroub, *Ultrason. Sonochem.* **2016**, *32*, 37–43.

29. K. Peng, F. Chen, X. She, C. Yang, Y. Cui, X. Pan, *Tetrahedron Lett.* **2005**, *46*, 1217–1220.
30. N. N. Mahamuni, P. R. Gogate, A. B. Pandit, *Ind. Eng. Chem. Res.* **2006**, *45*, 8829–8836.
31. S. Brochette-Lemoine, S. Trombotto, D. Joannard, G. Descotes, A. Bouchu, Y. Queneau, *Ultrason. Sonochem.* **2000**, *7*, 157–161.
32. D. Rinsant, G. Chatel, F. Jérôme, *ChemCatChem* **2014**, *6*, 3355–3359.
33. M. J. Climent, A. Corma, S. Iborra, *Green Chem.* **2011**, *13*, 520–540.
34. B. M. Despeyroux, K. Deller, E. Peldszus, In: *New Developments in Selective Oxidation* (Eds.: G. Centi, F. Trifiro), Elsevier, Amsterdam, 1990, 159.
35. R. S. Disselkamp, K. M. Judd, T. R. Hart, C. H. F. Peden, G. J. Posakony, L. J. Bond, *J. Catal.* **2004**, *221*, 347–353.
36. B. Tripathi, L. Paniwnyk, N. Cherkasov, A. O. Ibhadon, T. Lana-Villarreal, R. Gómez, *Ultrason. Sonochem.* **2015**, *26*, 445–451.
37. V. Hessel, G. Cravotto, P. Fitzpatrick, B. S. Patil, J. Lang, W. Bonrath, *Chem. Eng. Process.* **2013**, *71*, 19–30.
38. R. S. Disselkamp, S. M. Chajkowski, K. R. Boyles, T. R. Hart, C. H. Peden, *Cavitating Ultrasound Hydrogenation of Water-Soluble Olefins Employed Inert Dopants, Studies of Activity, Selectivity and Reaction Mechanisms*, CRC Press, Taylor & Francis Group, Boca Raton, Florida, USA, 2006.
39. A. Khorshidi, *Chin. J. Catal.* **2016**, *37*, 153–158.
40. J. Safari, L. Javadian, *Ultrason. Sonochem.* **2015**, *22*, 341–348.
41. M. Nasrollahzadeh, A. Ehsani, A. Rostami-Vartouni, *Ultrason. Sonochem.* **2014**, *21*, 275–282.
42. G. Cravotto, P. Cintas, *Chem. Soc. Rev.* **2006**, *35*, 180–196.
43. K. Suslick, G. J. Prince, *Annu. Rev. Mater. Sci.* **1999**, *29*, 295–326.
44. H. Xu, B. W. Zeiger, K. Suslick, *Chem. Soc. Rev.* **2013**, *42*, 2555–2567.
45. P. Cintas, J.-L. Luche, In: *Synthetic Organic Sonochemistry*, (Ed.: J.-L. Luche), Plenum Press, New York, 1998, 167–234.
46. G. Cravotto, E. C. Gaudino, P. Cintas, *Chem. Soc. Rev.* **2013**, *42*, 7521–7534.
47. D. G. Shchukin, E. Skorb, V. Belova, H. Möhwald, *Adv. Mater.* **2011**, *23*, 1922–1934.
48. S. Verdan, G. Burato, M. Comet, L. Reinert, H. Fuzellier, *Ultrason. Sonochem.* **2003**, *10*, 291–295.
49. A. Amanov, O. V. Penkov, Y.-S. Pyun, D.-E. Kim, *Tribol. Int.* **2012**, *54*, 106–113.

50. Y. Mizukoshi, Y. Makise, T. Shuto, J. Hu, A. Tominaga, S. Shironita, S. Tanabe, *Ultrason. Sonochem.* **2007**, *14*, 387–392.

51. G. Chatel, L. Novikova, S. Petit, *Appl. Clay Sci.* **2016**, *119*, 193–201.

52. A. N. Nguyen, L. Reinert, J.-M. Lévêque, A. Beziat, P. Dehaudt, J.-F. Juliaa, L. Duclaux, *Appl. Clay Sci.* **2016**, *72*, 9–17.

53. V. Belova, H. Möhwald, D. G. Shchukin, *Langmuir* **2008**, *24*, 9747–9753.

54. J. Yu, M. Zhou, B. Cheng, H. Yu, X. Zhao, *J. Mol. Catal. A: Chem.* **2005**, *227*, 75–80.

55. Z.-Q. Li, L.-G. Qiu, T. Xu, Y. Wu, W. Wang, Z.-Y. Wu, X. Jiang, *Mater. Lett.* **2009**, *63*, 78–80.

56. S. Fu, G. Yang, Y. Zhou, H.-B. Pan, C. M. Wai, D. Du, Y. Lin, *RSC Adv.* **2015**, *5*, 32685–32689.

57. S. Seok, M. A. Hussain, K. J. Park, J. W. Kim, D. H. Kim, *Ultrason. Sonochem.* **2016**, *28*, 178–184.

58. S. Allahyari, M. Haghighi, A. Ebadi, S. Hosseinzadeh, *Energy Convers. Manage.* **2014**, *83*, 212–222.

59. H. M. Xiong, D. G. Shchukin, H. Mohwald, Y. Xu, Y. Y. Xia, *Angew. Chem. Int. Ed.* **2009**, *48*, 2727–2731.

60. H. X. Xu, K. S. Suslick, *J. Am. Chem. Soc.* **2011**, *133*, 9148–9151.

61. K. S. Suslick, M. W. Grinstaff, *J. Am. Chem. Soc.* **1990**, *112*, 7807–7809.

62. M. Bradley, M. Ashokkumar, F. Grieser, *J. Am. Chem. Soc.* **2003**, *125*, 525–529.

63. B. M. Teo, F. Grieser, M. Ashokkumar, Applications of ultrasound to polymer synthesis, In: *Handbook on Applications of Ultrasound: Sonochemistry for Sustainability* (Eds.: D. Chen, S. K. Sharma, A. Mudhoo), CRC Press, Taylor & Francis Group, Boca Raton, FL, USA, 2011, 475–500.

64. G. J. Price, Polymer sonochemistry: Controlling the structure and properties of macromolecules, In: *Sonochemistry and Sonoluminescence* (Eds.: L. A. Crum, T. J. Mason, J. L. Reisse, K. S. Suslick) NATO ASI Series, Vol. 524, 1999, 321–343.

65. G. J. Price, P. F. Smith, P. J. West, *Ultrasonics* **1991**, *29*, 166–170.

66. G. J. Price, P. J. West, P. F. Smith, *Ultrason. Sonochem.* **1994**, *1*, S51–S57.

67. G. Cravotto, P. Cintas, *Chem. Soc. Rev.* **2009**, *38*, 2684–2697.

68. G. Kaupp, *J. Phys. Org. Chem.* **2008**, *21*, 630–643.

69. M. M. Caruso, D. A. Davis, Q. Shen, S. A. Odom, N. R. Sottos, S. R. White, J. S. Moore, *Chem. Rev.* **2009**, *109*, 5755–5798.

70. A. V. Mohod, P. R. Gogate, *Ultrason. Sonochem.* **2011**, *18*, 727–734.

71. R. Czechowska-Biskup, B. Rokita, S. Lotfy, P. Ulanski, J. M. Rosiak, *Carbohydr. Polym.* **2005**, *60*, 175–184.

72. P. R. Gogate, A. L. Prajapat, *Ultrason. Sonochem.* **2015**, *27*, 480–494.
73. S. Koda, K. Taguchi, K. Futamura, *Ultrason. Sonochem.* **2011**, *18*, 276–281.
74. M. A. K. Mostafa, *J. Polym. Sci.* **1958**, *28*, 519–536.
75. M. Schaefer, B. Icli, C. Weder, M. Lattuada, A. F. M. Kilbinger, Y. C. Simon, *Macromol.* **2016**, *49*, 1630–1636.
76. P. A. May, N. F. Munaretto, M. B. Hamoy, M. J. Robb, J. S. Moores, *ACS MacroLett.* **2016**, *5*, 177–180.
77. M. Behrens, A. K. Datye, *Catalysis for the Conversion of Biomass and Its Derivatives*, epubli GmbH, 2013.
78. K. Ullah, V. K. Sharma, S. Dhingra, G. Braccio, M. Ahmad, S. Sofia, *Renewable Sustainable Energy Rev.* **2015**, *51*, 682–698.
79. M. G. Adsul, M. S. Singhvi, S. A. Gaikaiwari, D. V. Gokhale, *Bioresour. Technol.* **2011**, *6*, 4304–4312.
80. R. Schlogl, Biomass conversion to chemicals, *In*: *Chemical Energy Storage*, De Gruter GmbH, Gottingen, 2012, *28*, 87–108.
81. P. Gallezot, *Chem. Soc. Rev.* **2012**, *41*, 1538–1558.
82. F. H. Isikgor, C. R. Becer, *Polym. Chem.* **2015**, *6*, 4497–4559.
83. T. Jain, J. V. Gerpen, A. McDonald, *J. Biofuels* **2010**, *1*, 109–114.
84. O. J. Sanchez, S. Montoya, *Production of Bioethanol from Biomass: An Overview*, *Biofuel Technologies*, Springer-Verlag, Berlin, Heidelberg, 2013, 397–441.
85. S. Nandaa, R. Azargoharb, A. K. Dalaib, J. A. Kozinski, *Renewable Sustainable Energy Rev.* **2015**, *50*, 925–941.
86. G.-Q. Chen, M. K. Patel, *Chem. Rev.* **2012**, *112*, 2082–2099.
87. Y. Habibi, L. A. Lucia, O. J. Rojas, *Chem. Rev.* **2010**, *110*, 3479–3500.
88. T. Gomiero, M. G. Paoletti, D. Pimentel, *J. Agric. Environ. Ethics* **2010**, *23*, 403–434.
89. A. Garcia, M. Gonzalez Alriols, R. Llano-Ponte, J. Labidi, *Bioresour. Technol.* **2011**, *102*, 6326–6330.
90. P. R. Gogate, M. K. Abhijeet M. Kabadi, *Biochem. Eng. J.* **2009**, *44*, 60–72.
91. E. V. Rokhina, P. Lens, J. Virkutyte, *Trends Biotechnol.* **2009**, *5*, 298–306.
92. J. Luo, Z. Fang, R. L. Smith Jr, Prog. *Energy Combust. Sci.* **2014**, *41*, 56–93.
93. P. Cintas, S. Mantegna, E. C. Gaudino, G. Cravotto, *Ultrason. Sonochem.* **2010**, *17*, 985–989.
94. M. J. Bussemaker, D. Zhang, *Ind. Eng. Chem. Res.* **2013**, *52*, 3563–3580.
95. F. Cheng, H. Wang, G. Chatel, G. Gurau, R. D. Rogers, *Bioresour. Technol.* **2014**, *164*, 394–401.
96. S. Singh, S. T. P. Bharadwaja, P. K. Yadav, V. S. Moholkar, A. Goyal, *Ind. Eng. Chem. Res.* **2014**, *53*, 14241–14252.

97. Y. Li, A. S. Fabiano-Tixier, V. Tomao, G. Cravotto, F. Chemat, *Ultrason. Sonochem.* **2013**, *20*, 12–18.

98. N. Kardos, J. L. Luche, *Carbohydr. Res.* **2001**, *332*, 115–131.

99. B. Karki, D. Maurer, S. Jung, *Bioresour. Technol.* **2011**, *102*, 6522–6528.

100. A. Iskalieva, B. M. Yimmou, P. R. Gogate, M. Horwath, P. G. Horvath, L. Csoka, *Ultrason. Sonochem.* **2012**, *19*, 984–993.

101. S. Y. Tang, M. Sivakumar, Ultrasound as a green processing technology for pretreatment and conversion of biomass into biofuels, In: *Production of Biofuels and Chemicals with Ultrasound* (Eds.: Z. Fang, R. L. Smith Jr. and X. Qi), Springer, Berlin, 2015, 189–207.

102. P. B. Subhedar, P. R. Gogate, *Ultrason. Sonochem.* **2014**, *21*, 216–225.

103. M. P. Badve, P. R. Gogate, A. B. Pandit, L. Csoka, *Ultrason. Sonochem.* **2014**, *21*, 162–168.

104. K. Suresh, A. Ranjan, S. Singh, V. S. Moholkar, *Ultrason. Sonochem.* **2014**, *21*, 200–207.

105. P. N. Patil, P. R. Gogate, L. Csoka, A. Dregelyi-Kiss, M. Horvath, *Ultrason. Sonochem.* **2016**, *30*, 79–86.

106. N. A. M. Liyakathali, P. D. Muley, G. Aita, D. Boldor, *Biores. Technol.* **2016**, *200*, 262–271.

107. B. Stefanovic, T. Rosenau, A. Potthast, *Carbohydr. Polym.* **2013**, *92*, 921–927.

108. S.-S. Wong, S. Kasapis, Y. M. Tan, *Carbohydr. Polym.* **2009**, *77*, 280–287.

109. K. Yasuda, D. Kato, Z. Xu, M. Sakka, K. Sakka, *Jpn. J. Appl. Phys.* **2010**, *49*, 07HE08.

110. M. J. Bussemaker, F. Xu, D. Zhang, *Bioresour. Technol.* **2013**, *148*, 15–23.

111. R. Velmurugan, K. Muthukumar, *Bioresour. Technol.* **2011**, *102*, 7119–7123.

112. Q. Zhang, M. Benoit, K. De Oliveira Vigier, J. Barrault, G. Jégou, M. Philippe, F. Jérôme, *Green Chem.* **2013**, *15*, 963–969.

113. C.-Y. Yang, I.-C. Sheih, T. J. Fang, *Ultrason. Sonochem.* **2012**, *19*, 687–691.

114. P. B. Subhedar, P. R. Gogate, *Ind. Eng. Chem. Res.* **2013**, *52*, 11816–11828.

115. G. Chatel, K. De Oliveira Vigier, F. Jérôme, *ChemSusChem* **2014**, *7*, 2774–2787.

116. B. Kwiatkowska, J. Bennnett, J. Akunna, G. M. Walker, D. H. Bremner, *Biotechnol. Adv.* **2011**, *29*, 768–780.

117. V. Yachmenev, B. Condon, T. Klasson, A. Lambert, *J. Biobased Mater. Bioenergy* **2009**, *3*, 25–31.

118. T. Wells Jr., M. Kosa, A. J. Ragauskas, *Ultrason. Sonochem.* **2013**, *20*, 1463–1469.

119. F. Napoly, N. Kardos, L. Jean-Gérard, C. Goux-Henry, B. Andrioletti, M. Draye, *Ind. Eng. Chem. Res.* **2015**, *54*, 6046–6051.

120. R. Behling, S. Valange, G. Chatel, *Green Chem.* **2016**, *18*, 1839–1854.

121. K. Ninomiya, H. Takamatsu, A. Onishi, K. Takahashi, N. Shimizu, *Ultrason. Sonochem.* **2013**, *20*, 1092–1097.

122. M. Kunaver, E. Jasiukaityte, N. Cuk, *Bioresour. Technol.* **2012**, *103*, 360–366.

123. L. E. Shaw, D. Lee, *Ultrason. Sonochem.* **2009**, *16*, 321–324.

124. S. Brochette-Lemoine, D. Joannard, G. Descotes, A. Bouchu, Y. Queneau, *J. Mol. Catal. A: Chem.* **1999**, *150*, 31–36.

125. S. Brochette-Lemoine, S. Trombotto, D. Joannard, G. Descotes, A. Bouchu, Y. Queneau, *Ultrason. Sonochem.* **2000**, *7*, 157–161.

126. S. Lemoine, C. Thomazeau, D. Joannard, S. Trombotto, G. Descotes, A. Bouchu, Y. Queneau, *Carbohydr. Res.* **2000**, *326*, 176–184.

127. Z.-Y. Qin, G.-L. Tong, Y. C. Frank, J.-C. Zhou, *Bioresour.* **2011**, *6*, 1136–1146.

128. S. P. Mishra, J. Thirree, A.-S. Manent, B. Chabot, C. Daneault, *Bioresour.* **2011**, *6*, 121–143.

129. M. Paquin, E. Loranger, V. Hannaux, B. Chabot, C. Daneault, *Ultrason. Sonochem.* **2013**, *20*, 103–108.

130. P. Bujak, P. Bartczak, J. Polanski, *J. Catal.* **2012**, *295*, 15–21.

131. B. Toukoniitty, J. Kuusisto, J.-P. Mikkola, T. Salmi, D. Y. Murzin, *Ind. Eng. Chem. Res.* **2005**, *44*, 9370–9375.

132. M. Sujka, J. Jamroz, *Food Hydrocolloids* **2013**, *31*, 413–419.

133. J. Zhu, L. Li, L. Chen, X. Li, *Food Hydrocolloids* **2012**, *29*, 116–122.

134. B. K. Tiwari, *Trends Anal. Chem.* **2015**, *71*, 100–109.

135. S. R. Shirsath, S. H. Sonawane, P. R. Gogate, *Chem. Eng. Process.* **2012**, *53*, 10–23.

136. C. Chen, L.-J. You, A. M. Abbasi, X. Fu, R. H. Liu, *Carbohydr. Polym.* **2015**, *130*, 122–132.

137. A.-G. Sicaire, M. A. Vian, F. Fine, P. Carré, S. Tostain, F. Chemat, *Ultrason. Sonochem.* **2016**, *31*, 319–329.

138. S. U. Kadam, B. K. Tiwari, C. Alvarez, C. P. O'Donnell, *Trends Food Sci. Technol.* **2015**, *46*, 60–67.

139. M. Yolmeh, M. B. Habini Najafi, R. Farhoosh, *Food Chem.* **2014**, *155*, 319–324.

140. F. J. Barba, Z. Zhu, M. Koubaa, A. S. Sant'Ana, V. Orlien, *Trends Food Sci. Technol.* **2016**, *49*, 96–109.

141. B. Pesic, T. Zhou, *Metall. Trans. B* **1992**, *23*, 13–22.

142. V. C. D. Peronico, J. L. Raposo Jr., *Food Chem.* **2016**, *196*, 1287–1292.

143. M.-F. Li, S.-N. Sun, F. Xu, R.-C. Sun, *Ultrason. Sonochem.* **2012**, *19*, 243–249.

144. J. X. Sun, X. F. Sun, H. Zhao, R. C. Sun, *Polym. Degrad. Stab.* **2004**, *84*, 331–339.

145. B. Karki, B. P. Lamsal, S. Jung, J. van Leeuwen, A. L. Pometto IIIe, D. Grewell, S. K. Khanal, *J. Food. Eng.* **2010**, *96*, 270–278.

146. R. Minjares-Fuentes, A. Femenia, M. C. Garau, M. G. Candelas-Cadillo, S. Simal, *Carbohydr. Polym.* **2015**, *23*, 148–155.

147. B. Yang, M. Zhao, J. Shi, N. Yang, Y. Jiang, *Food. Chem.* **2008**, *106*, 685–690.

148. B. Yang, Y. Jiang, M. Zhao, J. Shi, L. Wang, *Polym. Degrad. Stab.* **2008**, *93*, 268–272.

149. B. Yang, M. Zhao, Y. Jiang, *Food. Chem.* **2008**, *110*, 294–300.

150. S. Rodrigues, G. A. S. Pinto, *J. Food Eng.* **2007**, *80*, 869–872.

151. S. Nitayavardhana, P. Shrestha, M. L. Rasmussen, B. P. Lamsal, J. van Leeuwen, S. K. Khanal, *Bioresour. Technol.* **2010**, *101*, 2741–2747.

152. G. Zhang, L. He, M. Hu, *Innovative Food Sci. Emerg. Technol.* **2011**, *12*, 18–25.

153. T.-Q. Yuan, F. Xu, J. He, R.-C. Sun, *Biotechnol. Adv.* **2010**, *28*, 583–593.

154. R. C. Sun, X. F. Sun, X. H. Ma, *Ultrason. Sonochem.* **2002**, *9*, 95–101.

155. N. Rombaut, A.-S. Tixier, A. Bily, F. Chemat, *Biofuels, Bioprod. Bioref.* **2014**, *8*, 530–544.

156. D. Pingret, A.-S. Fabiano-Tixier, C. Le Bourvellec, C. M. G. C. Renard, F. Chemat, *J. Food. Eng.* **2012**, *111*, 73–81.

157. G. Chatel, D. R. MacFarlane, *Chem. Soc. Rev.* **2014**, *43*, 8132–8149.

158. P. G. Jessop, *Green Chem.* **2011**, *13*, 1391–1398.

159. C. A. Angell, Y. Ansari and Z. Zhao, *Faraday Discuss.* **2012**, *154*, 9–27.

160. P. Wasserscheid and T. Welton, *Ionic Liquids in Synthesis*, 2nd Edn., Wiley-VCH Verlag GmbH & Co. KGaA, Weinheim, 2008.

161. N. V. Plechkova, R. D. Rogers and K. R. Seddon, *Ionic Liquids: From Knowledge to Application*, ACS Symposium Series, 2010.

162. S. Aparicio, M. Atilhan and F. Karadas, *Ind. Eng. Chem. Res.* **2010**, *49*, 9580–9595.

163. J. P. Hallett and T. Welton, *Chem. Rev.* **2011**, *111*, 3508–3576.

164. D. R. MacFarlane, N. Tachikawa, M. Forsyth, J. M. Pringle, P. C. Howlett, G. D. Elliott, J. H. Davis Jr., M. Watanabe, P. Simon, C. A. Angell, *Energy Environ. Sci.* **2014**, *7*, 232–250.

165. Q. Zhang and J. M. Shreeve, *Chem. Eur. J.* **2013**, *19*, 15446–15451.

166. R. Peng, Y. Wang, W. Tang, Y. Yang, X. Xie, *Polymers* **2013**, *5*, 847–872.

167. G. Chatel, R. D. Rogers, *ACS Sustainable Chem. Eng.* **2014**, *2*, 322–339.
168. A. Brandt, J. Gräsvik, J. P. Hallett, T. Welton, *Green Chem.* **2013**, *15*, 550–583.
169. T. D. Ho, C. Zhang, L. W. Hantao, J. L. Anderson, *Anal. Chem.* **2014**, *86*, 262–285.
170. M. J. Trujillo-Rodríguez, P. Rocío-Bautista, V. Pino, A. M. Afonso, *TrAC, Trends Anal. Chem.* **2013**, *51*, 87–106.
171. J. L. Shamshina, S. P. Kelley, G. Gurau, R. D. Rogers, *Nature* **2015**, *528*, 188–189.
172. A. A. J. Torreiro, *Electrochemistry in Ionic Liquids, Vol. 1: Fundamentals*, Springer International Publishing, Switzerland, 2015, 625.
173. J. D. Holbrey, K. R. Seddon, *Clean Products and Processes* (Ed. T. Matsunaga), Vol. 1, Springer-Verlag, New York, 1999, 223.
174. R. R. Deshmukh, R. Rajagopal, K. V. Srinivasan, *Chem. Commun.* **2001**, 1544–1545.
175. R. Rajagopal, D. V. Jarikote, K. V. Srinivasan, *Chem. Commun.* **2002**, 616–617.
176. R. Rajagopal, K. V. Srinivasan, *Ultrason. Sonochem.* **2003**, *10*, 41–43.
177. A. R. Gholap, K. Venkatesan, T. Daniel, R. J. Lahoti, K. V. Srinivasan, *Green Chem.* **2003**, *5*, 693–696.
178. A. R. Gholap, K. Venkatesan, T. Daniel, R. J. Lahoti, K. V. Srinivasan, *Green Chem.* **2004**, *6*, 147–150.
179. A. R. Gholap, K. Venkatesan, R. Pasricha, T. Daniel, R. J. Lahoti, K. V. Srinivasan, *J. Org. Chem.* **2005**, *70*, 4869–4872.
180. K. Venkatesan, S. S. Pujari, R. J. Lahoti, K. V. Srinivasan, *Ultrason. Sonochem.* **2008**, *15*, 548–553.
181. W. Pei, C. Shen, *Chin. Chem. Lett.* **2006**, *17*, 1534–1536.
182. W. Bonrath, U. Létinois, T. Netscher, J. Schütz, Mizoroki–Heck reactions: Modern solvent systems and reaction techniques, In: *The Mizoroki–Heck Reaction* (Ed.: M. Oestreich), John Wiley & Sons, Ltd, Chichester, UK, 2009.
183. L. Bai, J.-X. Wang, *Curr. Org. Chem.* **2005**, *9*, 535–553.
184. S. Zhao, X. Wang, L. Zhang, *RSC Adv.* **2013**, *3*, 11691–11696.
185. S. Zhao, E. Zhao, P. Shen, M. Zhao, J. Sun, *Ultrason. Sonochem.* **2008**, *15*, 955–959.
186. R. S. Porto, G. W. Amarante, M. Cavallaro, R. J. Poppi, F. Coelho, *Tetrahedron Lett.* **2009**, *50*, 1184–1187.
187. J. R. Harjani, T. J. Abraham, A. T. Gomez, M. T. Garcia, R. D. Singer, P. J. Scammells, *Green Chem.* **2010**, *12*, 650–655.

188. F. D'Anna, S. Marullo, P. Vitale, R. Noto, *Ultrason. Sonochem.* **2012**, *19*, 136–142.

189. N. Yadav, M. K. Hussain, M. I. Ansari, P. K. Gupta, K. Hajela, *RSC Adv.* **2013**, *3*, 540–544.

190. S. Narayanaperumal, R. C. Da Silva, K. S. Feu, A. F. De La Torre, A. G. Corrêa, M. W. Paixão, *Ultrason. Sonochem.* **2013**, *20*, 793–798.

191. S. Rostamnia, M. Amini, *Chem. Pap.* **2014**, *68*, 834–837.

192. J.-M. Lévêque, G. Cravotto, *CHIMIA Int. J. Chem.* **2006**, *60*, 313–320.

193. J. Estager, Integrating ultrasound with other green technologies: Towards sustainable chemistry, In: *Handbook on Applications of Ultrasound: Sonochemistry for Sustainability* (Eds.: D. Chen, S. K. Sharma and A. Mudhoo), CRC Press, USA, 2011, 675–692.

194. G. Chatel, C. Goux-Henry, N. Kardos, J. Suptil, B. Andrioletti, M. Draye, *Ultrason. Sonochem.* **2012**, *19*, 390–394.

195. G. Chatel, C. Goux-Henry, A. Mirabaud, T. Rossi, N. Kardos, B. Andrioletti, M. Draye, *J. Catal.* **2012**, *291*, 127–132.

196. M. Mamaghani, M. Pourranjbar, R. H. Nia, *J. Sulfur. Chem.* **2014**, *1*, 1–6.

197. J. Wang, Y. Zong, R. Fu, Y. Niu, G. Yue, Z. Quan, X. Wang,Y. Pan, *Ultrason. Sonochem.* **2014**, *21*, 29–34.

198. D. Li, H. Zang, C. Wu, N. Yu, *Ultrason. Sonochem.* **2013**, *20*, 1144–1148.

199. Suresh, J. S. Sandhu, *Org. Med. Chem. Lett.* **2013**, *3*, 2–8.

200. H. Qian, Y. Wang, D. Liu, *Ind. Eng. Chem. Res.* **2013**, *52*, 13272–13275.

201. J. Estager, J.-M. Lévêque, R. Turgis, M. Draye, *Tetrahedron Lett.* **2007**, *5*, 755–759.

202 Z. Yinghuai, S. Bahnmueller, N. S. Hosmane, J. A. Maguirey, *Chem. Lett.* **2003**, *32*, 730–731.

203. V. V. Namboodiri, R. S. Varma, *Org. Lett.* **2002**, *4*, 3161–3163.

204. J.-M. Lévêque, J.-L. Luche, C. Pétrier, R. Roux, W. Bonrath, *Green Chem.* **2002**, *4*, 357–360.

205. R. S. Varma, *J. Chem.* **2006**, *45B*, 2305–2312.

206. J.-M. Lévêque, S. Desset, J. Suptil, C. Fachinger, M. Draye, W. Bonrath, G. Cravotto, *Ultrason. Sonochem.* **2006**, *13*, 189–193.

207. G. Cravotto, E. C. Gaudino, L. Boffa, J.-M. Lévêque, J. Estager, W. Bonrath, *Molecules* **2008**, *13*, 149–156.

208. S. Zhao, E. Zhao, P. Shen, M. Zhao, J. Sun, *Ultrason. Sonochem.* **2008**, *15*, 955–959.

209. W. Li, Q. Lin, L. Ma, *Ultrason. Sonochem.* **2010**, *17*, 752–755.

210. M. Messali, M. A. M. Asiri, *J. Mater. Environ. Sci.* **2013**, *4*, 770–785.

211. M. Messali, *Arabian J. Chem.* **2014**, *7*, 63–70.

212. G. Zbancioc, I. I. Mangalagiu, C. Moldoveanu, *Ultrason. Sonochem.* **2015**, *23*, 376–384.

213. R. S. Varma, Expeditious synthesis of ionic liquids using ultrasound and microwave irradiation, In: *Ionic Liquids as Green Solvents* (Eds.: R. D. Rogers, K. R. Seddon), ACS Symposium Series, American Chemical Society, Washington, 2003, Vol. 856, 782–92.

214. R. S. Varma, *Green Chem. Lett. Rev.* **2007**, *1*, 37–45.

215. J.-M. Lévêque, J. Estager, M. Draye, G. Cravotto, L. Boffa, W. Bonrath, *Monatsh. Chem.* **2007**, *138*, 1103–1113.

216. M. Deetlefs, K. R. Seddon, *Green Chem.* **2010**, *12*, 17–30.

217. J. D. Oxley, T. Prozorov, K. S. Suslick, *J. Am. Chem. Soc.* **2003**, *125*, 11138–11139.

218. G. Chatel, R. Pflieger, E. Naffrechoux, S. I. Nikitenko, J. Suptil, C. Goux-Henry, N. Kardos, B. Andrioletti, M. Draye, *ACS Sustainable Chem. Eng.* **2013**, *1*, 137–143.

219. Z. Yinghuai, S. Bahnmueller, C. Chibun, K. Carpenter, N. S. Hosmane, J. A. Maguire, *Tetrahedron Lett.* **2003**, *44*, 5473–5476.

220. Z. Yinghuai, *J. Phys. Chem. Solids* **2004**, *65*, 349–353.

221. X. Wang, P. F. Fulvio, G. A. Baker, G. M. Veith, R. R. Unocic, S. M. Mahurin, M. Chib, S. Dai, *Chem. Commun.* **2010**, *46*, 4487–4489.

222. J.-J. Shi, J. J. Zhu, *Electrochim. Acta* **2011**, *56*, 6008–6013.

223. C. Z. Zhu, S. J. Guo, Z. Y. Zhai, S. J. Dong, *Langmuir* **2010**, *26*, 7614–7618.

224. V. S. Nalajala, V. S. Moholkar, *Ultrason. Sonochem.* **2011**, *18*, 345–355.

225. W. Xiao, Z. Sun, S. Chen, H. Zhang, Y. Zhao, C. Huang, Z. Liu, *RSC Adv.* **2012**, *2*, 8189–8193.

226. B.-H. Mao, C.-H. Liu, X. Gao, R. Chang, Z. Liu, S.-D. Wang, *Appl. Surf. Sci.* **2013**, *283*, 1076–1079.

227. K. L. Luska, A. Moores, *Green Chem.* **2012**, *14*, 1736–1742.

228. K. L. Luska, A. Moores, *ChemCatChem.* **2012**, *4*, 1534–1546.

229. K. L. Luska, A. Moores, *Adv. Synth. Catal.* **2011**, *353*, 3167–3177.

230. A. Wittmar, D. Ruiz-Abad, M. Ulbricht, *J. Nanopart. Res.* **2012**, *14*, 651–661.

231. K. V. P. M. Shafi, A. Ulman, A. Dyal, X. Yan, N.-L. Yang, C. Estournès, L. Fournès, A. Wattiaux, H. White, M. Rafailovich, *Chem. Mater.* **2002**, *14*, 1778–1787.

232. J. P. Bazureau, M. Draye, *Ultrasound and Microwaves: Recent Advances in Organic Chemistry*, Research Signpost, 2011.

233. K. S. Suslick, D. A. Hammerton, D. E. Cline, *J. Am. Chem. Soc.* **1986**, *108*, 5641–5645.

234. Y. Jin, P. Wang, D. Yin, J. Liu, L. Qin, N. Yu, G. Xie, B. Li, *Colloids Surf. A* **2007**, *302*, 366–370.

235. K.-S. Kim, D. Demberelnyamba, H. Lee, *Langmuir* **2004**, *20*, 556–560.

236. V. Taghvaei, A. Habibi-Yangjeh, M. Behboudnia, *Powder Technol.* **2009**, *195*, 63–67.

237. M. Behboudnia, A. Habibi-Yangjeh, Y. Jafari-Tarzanag, A. Khodayari, *J. Optoelectron. Adv. Mater.* **2009**, *11*, 134–139.

238. M. Behboudnia, A. Habibi-Yangjeh, Y. Jafari-Tarzanag, A. Khodayari, *Bull. Korean Chem. Soc.* **2008**, *29*, 53–56.

239. M. Behboudnia, A. Habibi-Yangjeh, Y. Jafari-Tarzanag, A. Khodayari, *J. Phys. Chem. Solids* **2010**, *71*, 1393–1397.

240. N. Bouropoulos, *Sci. Adv. Mater.* **2013**, *5*, 46–50.

241. T. Alammar, A. Birkner, O. Shekhah, A.-V. Mudring, *Mater. Chem. Phys.* **2010**, *120*, 109–113.

242. U. Sang Shin, H.-K. Hong, H.-W. Kim, M.-S. Gong, *Bull. Korean Chem. Soc.* **2011**, *32*, 1583–1586.

243. S. Zhang, Y. Zhang, Y. Wang, S. Liu, Y. Deng, *Phys. Chem. Chem. Phys.* **2012**, *14*, 5132–5138.

244. C. F. Poole, S. K. Poole, *J. Chromatogr. A* **2010**, *1217*, 2268–2286.

245. J. G. Huddleston, H. D. Willauer, R. P. Swatloski, A. E. Visser, R. D. Rogers, *Chem. Commun.* **1998**, 1765–1766.

246. X. Sun, H. Luo, S. Dai, *Chem. Rev.* **2012**, *112*, 2100–2128.

247. X. Cao, X. Ye, Y. Lu, Y. Yu, W. Mo, *Anal. Chim. Acta* **2009**, *640*, 47–51.

248. Q. Zhou, X. Zhang, J. Xiao, *J. Chromatogr. A* **2009**, *1216*, 4361–4365.

249. J. Abolhasani, M. Amjadi, J. Hassanzadeh, E. Ghorbani-Kalhor, *Anal. Lett.* **2014**, *47*, 1528–1540.

250. S.-W. He, C.-Y. Shen, X.-Q. Wei, M.-C. Jin, M.-Q. Cai, *Adv. Mater. Res.* **2013**, *726–731*, 74–80.

251. D. Han, K. H. Row, *J. Sci. Food. Agric.* **2011**, *91*, 2888–2892.

252. K. Wu, Q. Zhang, Q. Liu, F. Tang, Y. Long, S. Yao, *J. Sep. Science* **2009**, *32*, 4220–4226.

253. S. Dong, Q. Hu, Z. Yang, R. Liu, G. Huang, T. Huang, *Microchem. J.* **2013**, *110*, 221–226.

254. Y. Sun, W. Li, J. Wang, *J. Chromatogr. B* **2011**, *879*, 975–980.

255. E. Molaakbari, A. Mostafavi, D. Afzali, *J. Hazard. Mater.* **2011**, *185*, 647–652.

256. E. Stanisz, J. Werner, H. Matusiewicz, *Microchem. J.* **2013**, *110*, 28–35.

257. M. M. P. Vazquez, P. P. Vazquez, M. M. Galera, M. D. G. Garcia, A. Ucles, *J. Chromatograph. A* **2013**, *1291*, 19–21.

258. M. Asensio-Ramos, J. Hernandez-Borges, T. M. Borges-Miquel, M. A. Rodriguez-Delgado, *J. Chromatogr. A* **2011**, *1218*, 4808–4816.
259. K. S. Suslick, S. J. Doktycz, E. B. Flint, *Ultrason.* **1990**, *5*, 280–290.
260. L. A. Crum, T. J. Mason, J. L. Reisse, K. S. Suslick, *Sonochemistry and Sonoluminescence*, NATO ASI Series, Washington, USA, 1998.
261. D. J. Flannigan, S. D. Hopkins, K. S. Suslick, *J. Organomet. Chem.* **2005**, *690*, 3513–3517.
262. K. S. Suslick, D. A. Hammerton, J. R. E. Cline, *J. Am. Chem. Soc.* **1986**, *108*, 5641–5642.
263. P. M. Kanthale, A. Brotchie, F. Grieser, M. Ashokkumar, *Ultrason. Sonochem.* **2013**, *20*, 47–51.
264. G. Chatel, L. Leclerc, E. Naffrechoux, C. Bas, N. Kardos, C. Goux-Henry, B. Andrioletti, M. Draye, *J. Chem. Eng. Data* **2012**, *57*, 3385–3390.
265. X. Li, J. Zhao, Q. Li, L. Wang, S. C. Tsang, *Dalton Trans.* **2007**, 1875–1880.
266. F. Yang, L. Li, Q. Li, W. Tan, W. Liu, M. Xian, *Carbohydr. Polym.* **2010**, *81*, 311–316.
267. S. Ho Ha, N. M. Hiep, Y.-M. Koo, *Biotechnol. Bioprocess Eng.* **2010**, *15*, 126–130.
268. Z. Liu, L. Lu, *Adv. Mater. Res.* **2011**, *236–238*, 169–172.
269. P. Lozano, B. Bernal, I. Recio, M.-P. Belleville, *Green Chem.* **2012**, *14*, 2631–2637.
270. Y. Wang, Y. Pan, Z. Zhang, R. Sun, X. Fang, D. Yu, *Process Biochem.* **2012**, *47*, 976–982.
271. K. Ninomiyaa, A. Kohorib, M. Tatsumib, K. Osawab, T. Endob, R. Kakuchib, C. Oginoc, N. Shimizua, K. Takahashi, *Bioresour. Technol.* **2015**, *176*, 169–174.
272. C.-Y. Yanga, T. J. Fang, *Process Biochem.* **2015**, *50*, 623–629.
273. K. Ninomiya, A. Ohta, S. Omote, C. Ogino, K. Takahashi, N. Shimizu, *Chem. Eng. J.* **2013**, *215–216*, 811–818.
274. S. H. Lee, H. M. Nguyen, Y.-M. Koo, S. H. Koo, S. H. Ha, *Process Biochem.* **2008**, *43*, 1009–1012.
275. J. Wang, S. Wang, Z. Li, S. Gua, X. Wu, F. Wu, *J. Mol. Catal. B, Enzym.* **2015**, *111*, 21–28.
276. C. Costa, M.-L. Doche, J.-Y. Hihn, I. Bisel, P. Moisy, J.-M. Lévêque, *Ultrasonics* **2010**, *50*, 323–328.
277. R. G. Compton, J. L. Hardcastle, J. del Campo, Encyclopedia of electrochemistry, In: *Instrumentation and Electroanalytical Chemistry*, (Eds.: Bard Stratmann, P. Unwin), Vol. 3, Ch. 2.9, Wiley-VCH, Weinheim, 2003.

278. F. Xiao, Z. Mo, F. Zhao, B. Zeng, *Electrochem. Commun.* **2008**, *10*, 1740–1743.
279. F. Xiao, F. Zhao, D. Mei, Z. Mo, B. Zeng, *Biosens. Bioelectron.* **2009**, *24*, 3481–3486.
280. F. Xiao, F. Zhao, Y. Zhang, G. Guo, B. Zeng, *J. Phys. Chem. C.* **2009**, *113*, 849–855.
281. F. Zhao, F. Xiao, B. Zeng, *Electrochem. Commun.* **2010**, *12*, 168–171.
282. Y. He, J. Zheng, S. Dong, *Analyst* **2012**, *137*, 4841–4848.
283. Y. He, J. Zheng, *Anal. Methods* **2013**, *5*, 767–772.
284. A. Loupy, *Microwaves in Organic Synthesis*, 2nd Edn., Wiley-VCH, Weinheim, 2006.
285. A. de la Hoz, A. Loupy, *Microwaves in Organic Synthesis*, Wiley-VCH, 2 Vol. Set, 2012, 1303.
286. A. de la Hoz, A. Díaz-Ortiz, A. Moreno, *Chem. Soc. Rev.* **2005**, *34*, 164–178.
287. M. B. Gawande, S. N. Shelke, R. Zboril, R. S. Varma, *Acc. Chem. Res.* **2014**, *47*, 1338–1348.
288. C. Leonelli, T. J. Mason, *Chem. Eng. Process.* **2010**, *49*, 885–900.
289. G. Cravotto, P. Cintas, *Chem. Eur. J.* **2007**, *13*, 1902–1909.
290. C. R. Strauss, R. W. Trainor, *Aust. J. Chem.* **1995**, *48*, 1665–1692.
291. H. M. Kingston, L. B. Jassie, In: *Introduction to Microwave Sample Preparation* (Eds.: H. M. Kingston, L. B. Jassie), American Chemical Society, Washington, DC, 1998, 231–233.
292. C. Buffa, G. Cravotto, G. Omiccioli Patent TO A-000766, 2006.
293. P. Cintas, S. Tagliapietra, M. Caporaso, S. Tabasso, G. Cravotto, *Ultrason. Sonochem.* **2015**, *25*, 8–16.
294. Z. Xua, Y. Yu, D. Fang, J. Xua J. Liang, L. Zhou, *Ultrason. Sonochem.* **2015**, *27*, 287–295.
295. S. Brundavanam, G. E. J. Poinern, D. Fawcett, *Am. J. Mater. Sci.* **2015**, *5*, 31–40.
296. G. Cravotto, L. Boffa, Preparation of nanomaterials under combined ultrasound/microwave irradiation, In: *Cavitation, A Novel Energy-Efficient Technique for the Generation of Nanomaterials* (Eds.: S. Manickam, M. Ashok kumar), Pan Stanford Publishing, Boco Raton, FL, 2014, 203–226.
297. G. Cravotto, M. Beggiato, A. Penoni, G. Palmisano, S. Tollari, J.-M. Lévêque, W. Bonrath, *Tetrahedron Lett.* **2005**, *46*, 2267–2271.
298. Y. Peng, G. Song, *Green Chem.* **2003**, *5*, 704–706.
299. M. Sacco, C. Charnay, F. de Angelis, M. Radoiu, F. Lamaty, J. Martinez, E. Colacino, *RSC Adv.* **2015**, *5*, 16878–16885.

300. F. Chemat, M. Poux, J.-L. Di Martino, J. Berlan, *J. Microwave Power Electromagn. Energy* **1996**, *31*, 19–22.

301. A. Lagha, S. Chemat, P. V. Bartels, F. Chemat, *Analusis* **1999**, *27*, 452–457.

302. S. Chemat, A. Lagha, H. A. Amar, F. Chemat, *Ultrason. Sonochem.* **2004**, *11*, 5–8.

303. Y. Peng, G. Song, *Green Chem.* **2001**, *3*, 302–304.

304. A. Hernoux-Villière, U. Lassi, T. Hu, A. Paquet, L. Rinaldi, G. Cravotto, S. Molina-Boisseau, M.-F. Marais, J.-M. Lévêque, *ACS Sustainable Chem. Eng.* **2013**, *1*, 995–1002.

305. A. Villière, G. Cravotto, R. Vibert, A. Perrier, U. Lassi, J.-M. Lévêque, Production of glucose from starch-based waste employing ultrasound and/or microwave irradiation, In: *Production of Biofuels and Chemicals with Ultrasound* (Eds.: Z. Fang, R. L. Smith, Jr., X. Qi), Springer: Netherlands, 2015, 289–315.

306. N. Morigushi, *J. Chem. Soc. Jpn* **1934**, *55*, 749–750.

307. J. González-García, M. D. Esclapez, P. Bonete, Y. V. Hernández, L. G. G. Garretón, V. Sáez, *Ultrasonics* **2010**, *50*, 318–322.

308. T. J. Mason, J. P. Lorimer, D. J. Walton, *Ultrasonics* **1990**, *28*, 333–337.

309. B. G. Pollet, S. S. Phull, *Recent Res. Develop. Electrochem.* **2001**, *4*, 55–78.

310. C. Brett, Sonoelectrochemistry, In: *Piezoelectric Transducers and Applications* (Ed.: A. A. Vives), Springer, Berlin Heidelberg, 2008, 399–411.

311. J. González-García, V. Sáez, M. D. Esclapez, P. Bonete, Y. Vargas, L. Gaete, *Physics Procedia* **2010**, *3*, 117–124.

312. B. Pollet, *Power Ultrasound in Electrochemistry: From Versatile Laboratory Tool to Engineering Solution*, Wiley-VCH, Beinheim, 2012, 364.

313. K. Zhang, S. Yao, G. Li, Y. Hu, *Nanoscale* **2015**, *7*, 2659–2666.

314. B. G. Pollet, J.-Y. Hihn, T. J. Mason, *Electrochim. Acta* **2008**, *53*, 4248–4256.

315. A. Mandroyan, M. Mourad-Mahmoud, M.-L. Doche, J.-Y. Hihn, *Ultrason. Sonochem.* **2014**, *21*, 2010–2019.

316. T. J. Mason, V. Saez Bernal, An introduction to sonoelectrochemistry, In: *Power Ultrasound in Electrochemistry: From Versatile Laboratory Tool to Engineering Solution* (Ed.: B. Pollet), Wiley-VCH, Beinheim, 2012, 21–44.

317. W. Lauterborn, *Appl. Sci. Res.* **1982**, *38*, 165–178.

318. B. Kwiatkowska, J. Bennnett, J. Akunna, G. M. Walker, D. H. Bremner, *Biotechnol. Adv.* **2011**, *29*, 768–780.

319. M. Wang, W. Yuan, X. Jiang, Y. Jing, Z. Wang, *Bioresour. Technol.* **2014**, *153*, 315–321.

320. A. Rehorek, M. Tauber, G Gübitz, *Ultrason. Sonochem.* **2004**, *11*, 177–182.
321. Y. G. Adewuyi, V. G. Deshmane, *Energy Fuels* **2015**, *29*, 4998–5006.
322. V. Yachmenev, B. Condon, T. Klasson, A. Lambert, *J. Biobased Mater. Bioenergy* **2009**, *3*, 25–31.
323. A. J. Borah, M. Agarwal, M. Poudyal, A. Goyal, V. S. Moholkar, *Bioresour. Technol.* **2016**, doi: 10.1016/j.biortech.2016.02.024.
324. M. Aliyu, M. J. Hepher, *Ultrason. Sonochem.* **2000**, *7*, 265–268.
325. C. Li, M. Yoshimoto, N. Tsukuda, K. Fukunaga, K. Nakao, *Biochem. Eng. J.* **2004**, *19*, 155–164.
326. C. Li, M. Yoshimoto, H. Ogata, N. Tsukuda, K. Fukunaga, K. Nakao, *Ultrason. Sonochem.* **2005**, *12*, 373–384.
327. Y. Zhang, E. Fu, J. Liang, *Chem. Eng. Technol.* **2008**, *31*, 1510–1515.
328. P. B. Subhedar, P. R. Gogate, *Ind. Eng. Chem. Res.* **2013**, *52*, 11816–11828.
329. P. B. Subhedar, P. R. Gogate, *J. Mol. Catal. B: Enzym.* **2014**, *101*, 180–114.
330. P. I. Neel, A. Gedanken, R. Schwarz, E. Sendersky, *Energy Fuels* **2012**, *26*, 2352–2356.
331. S. Singh, S. Sarma, M. Agarwal, A. Goyal, V. S. Moholkar, *Bioresour. Technol.* **2015**, *188*, 287–294.
332. R. Mawson, M. Gamage, N. S. Terefe, K. Knoerzer, Ultrasound in enzyme activation and inactivation, In: *Ultrasound Technologies for Food and Bioprocessing, Food Engineering*, 2011, 369–404.
333. Y. He, *Sonochemistry and Advances Oxidation Processes: Synthesis of Nanoparticles and Degradation of Organic Pollutants*, Thesis, School of Chemistry, The University of Melbourne, 2009.
334. T. Y. Wu, N. Guo, C. Y. The, J. X. W. Hay, *Advances in Ultrasound Technology for Environmental Remediation*, Springer, Netherlands, 2013.
335. S. Gligovski, R. Strekowski, S. Barbati, D. Vione, *Chem. Rev.* **2015**, *115*, 13051–13092.
336. R. Singla, F. Grieser, M. Ashokkumar, *Ultrason. Sonochem.* **2011**, *18*, 484–488.
337. Y. G. Adewuyi, *Environ. Sci. Technol.* **2005**, *39*, 3409–3420.
338. Y. G. Adewuyi, *Environ. Sci. Technol.* **2005**, *39*, 8557–8570.
339. C. Gong, J. Jiang, D. Li, S. Tian, *Sci. Rep.* **2015**, *5*, 11419.
340. P. Cintas, *Ultrason. Sonochem.* **2016**, *28*, 257–258.
341. T. J. Mason, *Ultrason. Sonochem.* **2007**, *14*, 476–483.

342. T. J. Mason, D. Peters, *Practical Sonochemistry. Power Ultrasound: Uses and Applications*, Ellis Horwood, Chichester, 2002.

343. T. Mason and J. Lorimer, *Applied Sonochemistry: Uses of Power Ultrasound in Chemistry and Processing*, Wiley-VCH, Beinheim, 2002.

344. J. P. Russell, M. Smith, *Advances in Sonochemistry* (Ed.: T. J. Mason), JAI Press, London, 1999, Vol. 5, 279–302.

Chapter 6

Conclusions and Outlook

This conclusive chapter will give some elements for future development of sonochemistry, especially for green chemistry applications. It starts with the limitations of sonochemistry, followed by a critical section to show the strong link between sonochemistry and green chemistry. To complete this chapter, an innovative section is proposed as a conclusion, since we invited some distinguished sonochemists from around the world to share their opinions on the past/present/future impacts of sonochemistry in terms of green chemistry: an exciting read to better know issues, challenges and trends in the field! The answers given here are personal points of view and referenced only to their authors. The content of this book does not represent researchers who contributed in this conclusive chapter.

1. Current Limitations of Sonochemistry

This section focuses on the limitations of sonochemistry. It is not to give the drawbacks of ultrasound applied to chemistry, but rather to highlight some neglected issues or improvements to develop in the field. It is also thanks to these limitations that the academic and industrial researchers still have many years of exciting discoveries in front of them.

Sonochemistry is a very broad subject, with many possible applications, but also a large amount of mysteries. Indeed, the interesting results obtained under ultrasound are not always explicable from a theoretical point of view. Sonochemical phenomena are not well known enough yet to understand what happens in every case. Mason concluded recently that strange and inexplicable aspects of sonochemistry are at the origin of several theories, for example the bubble collapse mechanisms that deserve to be extensively tested and demonstrated.[1] All scientists involved agreed that there were shock waves and jets generated in the bulk medium since this was evidenced by both physical effects and a picture of the bubble of cavitation.[2,3] Several **electrical theories** have been developed by Frenkel[4] and Margulis.[5] In addition, the **plasma theory** initially proposed by Lepoint *et al.*[6] is now discussed by Nikitenko *et al.* based on recent spectroscopic studies suggesting the non-equilibrium plasma formation inside the collapsing bubbles.[7] Hoffmann *et al.* also proposed that the cavitation bubble can generate **supercritical water**.[8] Even if it does not explain all the sonochemical phenomenon, the **hot-spot theory**, with a collapse of the cavitation bubble leading to intense conditions of temperatures and pressures, remains currently the most popular and is widely accepted in the community (see chapter entitled *Acoustic Cavitation*, page 13).[9,10] Thus, this lack of fundamental knowledge about sonochemical phenomena could limit the explanation of exciting results sometimes provided by ultrasound in chemistry.

For this reason, the **rigorous characterization** of sonochemical parameters (see chapter entitled *Ultrasonic Parameters Estimation*, page 27) is essential to understand the associated chemistry and to facilitate the comparison between each study reported in the literature. All the sonochemical parameters and experimental conditions have to be rigorously reported in the experimental part of publications. It is inconceivable to find articles without the details on the frequency, the power or the type of used reactors. A systematic comparison with corresponding silent conditions is also required to clearly highlight the ultrasonic effects.

A linked problem is the **lack of reproducibility** since many factors can affect the cavitation and the corresponding effects of ultrasound, such as the geometry of the reactor, the type of probe, the medium through which the waves are propagated, the concentration of dissolved gases, etc. Once again, it is very essential to describe with the maximum of detail the experimental conditions. The results obtained for the same reaction performed in different laboratories could be very different and even totally opposed. The important thing is to understand which parameter(s) is/are at the origin of these different reactivities in order to better understand the actions of ultrasound.

The **used equipment** can also limit the innovations in the area. In general, the ultrasonic field is not homogeneous in the whole medium. Once again, this inhomogeneity of the ultrasonic field could be problematic for both reproducibility and effective understanding of the phenomena occurring during ultrasonic irradiation.[11] The design of **new and specific sonochemical reactors** should be one of the key parameters to develop new processes at lab and semi-industrial scale.

Issues in **scaling up of sonoreactors** to meet industrial needs such as process efficiency and rates, energy conversion, high volume processes and others present a considerable challenge towards further development of sonochemistry in green chemistry.[12] Some design improvements to be investigated include transducer arrays and a larger exposed surface for ultrasound source, continuous flow reactor designs and stirring during sonication.

2. Sonochemistry: More than a Tool for Green Chemistry

In Chapter 5 of this handbook (*Applications in Green Chemistry*, page 59), new opportunities provided by sonochemistry in different areas of chemistry have been highlighted through several recent examples. Even if the mechanism studies are not always reported, many applications of ultrasound represent innovative and attractive advances from a green chemistry point of view. Indeed, as presented in Chapter 1 of this handbook (page 1), green chemistry is based on five basic concepts: (i) prevention, (ii) better use of the raw material, (iii) better waste management, (iv) energy savings and (v) use of solvents compatible with the environment.[13,14] When the experimental conditions are optimized, the use of ultrasound is, in the majority of cases, in favor of the 12 principles of green chemistry. With better yields and selectivities, reduced reaction times, new reactivity, use of water as solvent, increasing the reactivity of catalysts or optimized with catalysts, the sonochemical reactions are often greener than those performed under silent conditions.[15]

As a personal point of view, I am convinced and I want to convince all the chemists that sonochemistry and the use of ultrasound are not just a simple mixing tool. It is more than a cleaning bath that is placed somewhere at the bottom *of* the laboratory. Cravotto and Cintas used the term "***distinctive chemistry***" to speak about sonochemistry, making the connection between "*cavitational chemistry*" and "*mechanochemistry*", or in other words between chemical (formation or radical species) and physical effects (mechanical force) of sonochemistry.[16] I agree with that point: ultrasound irradiation is not a tool, but a real field of chemistry, with its specificities and its many mysteries that remain a challenge for researchers!

In their recent manuscript entitled "Misinterpretation of Green Chemistry" published in *Ultrasonics Sonochemistry*, Fegade and Trembly pointed out the misuse of the terms green/sustainable in the scientific literature.[17] In their paper, they proposed to educate readers on the importance of reactant and product toxicity information used in a **green** reaction. Cintas brought a response to this latter paper and some personal comments that I totally share, in his article entitled "Ultrasound and Green Chemistry — Further Comments" published in the same journal.[18] Indeed, considerable criticism and skepticism are necessary to appreciate the greenness of a reaction. In reality, it is essential to reason **on the overall process from raw materials to products and the costs of eliminating or recycling wastes**.

It is also important to link the developed sonochemical processes to the 12 principles of Green Engineering (Chapter 1, page 7), since the use of ultrasound at an industrial level in chemistry is clearly dependent on **(i) the scale-up, (ii) the energy consumption** and **(iii) the design of equipment**. In addition, these principles also take into account the process steps and afterlife of products.

These three key points are closely linked: no scale-up without specifically designed equipment, no scale-up without low energy consumption! Two strategies should be led in parallel: (1) the need for innovation on sonoreactors and transducers (design, optimization, etc.); (2) the need for sonochemists to systematically report an estimation of the dissipated ultrasonic power and energy costs, even if these data could also depend on reactor type and configuration. The **chemical dosimetry** is also an important parameter to systematically report and to identify the optimum power that should be applied to achieve the highest efficiency at an industrial scale (Chapter 3, page 27).[19] This rigor is important to better understand how ultrasound works, the reactions or the treatments under ultrasound, and how to move to higher scales. Microfluidics could also represent an innovative solution in terms of green chemistry, with the use of small quantities of reagents and solvents, and hence less waste, a precise control of reaction conditions, integration of functionality for process intensification, safer and often faster protocols, reliable scale-up and possibility of performing multiphase reactions.[20]

The choice of the reactors and the irradiation mode can greatly influence the reactivity under ultrasound. For example, the more systematic use of **tubular sonoreactors**, with diameter and length of the tube optimized for ensuring the proper transfer of energy, is encouraged since it represents an excellent approach to consider an industrial transposition by continuous flow irradiation. In addition, the walls of the tube become the ultrasonic emission surface and that is why the cavitation is mainly focused in the center of the tube, limiting the risk of corrosion. The development of continuous and circulating processes with an ultrasonic chamber is also encouraged. In the latter case, the irradiated volume and the irradiation time are two important parameters to consider.

Finally, **the fundamental studies** on sonochemistry, cavitation and ultrasonic effects are essential to better understand and know the systems. This exciting research on the mysteries around sonochemistry is also important for the sonochemists' community for future discussions, debates, proofs of concept and the scientific emulation of the field!

As a conclusion, sonochemistry has its part to play in the development of sustainable, green and eco-efficient processes to go ahead in terms of innovation, with the wildest results!

3. When Sonochemists Give Their Opinion...

Some famous sonochemists from around the world, with different sets of expertise and internationally known in their fields of research, accepted to give their personal points of view on what they think about the opportunities brought by sonochemistry for green chemistry. In other words, we asked them how sonochemistry contributed/contributes/will contribute to green chemistry according to them. A **great thanks to them** for sharing their opinion in this book to give a broader spectrum of future trends in the use of ultrasound in green chemistry. It is the best conclusion to complete this book.

> *The content of this book does not represent researchers who contributed in this chapter. The answers given here are personal points of view and commit only to their authors.*

Prof. Christos Argirusis

Laboratory of Inorganic Materials Technology (LIMT), School of Chemical Engineering, National Technical University of Athens (Greece)
I believe that sonochemistry/sonoelectrochemistry gives a tremendous opportunity in the field of catalyst preparation and ultrasound supported catalytic reactions. Using ultrasound with or without electrochemical reaction steps enables us to prepare catalysts with tailored properties. This leads to high conversion rates of the starting materials, to products with the desired functionality, helps to eliminate byproducts and waste which can be hazardous and in any case to reduced energy demand of the end-product due to avoiding cleaning and purification steps. A huge overall environmental and economic impact is expected using ultrasound in catalysis and catalyst preparation.

Prof. Muthupandian Ashokkumar

School of Chemistry, The University of Melbourne (Australia)
Sonochemistry offers "greener" reaction pathways. For example, sono-chemical reactions in general require lower temperatures, less amount of toxic solvents and shorter reaction times. All such reaction conditions lead to the generation of less waste and consumption of less energy providing better and greener environments. For example, the reaction time for enzyme catalyzed biofuel production could be reduced from 24 hours to just 1 hour and extraction of biofunctional compounds from plant materials could be achieved at lower reaction temperatures with fewer amounts of organic solvents.

Prof. Farid Chemat

GREEN Extraction Team, University of Avignon and the Vaucluse, INRA, UMR408 (France)

Ultrasound in food processing is an innovative, green and efficient concept to meet the challenges of the 21st Century protecting both consumers and environment, while in the meantime enhancing competitiveness of industries by becoming more economic, innovative and sustainable. Using ultrasound, food processes can now be completed in seconds or minutes with high reproducibility, reducing the processing cost, simplifying manipulation and workup, giving higher purity of the final product, eliminating post-treatment of waste water and consuming only a fraction of the time and energy normally needed for conventional processes. Several processes such as freezing, cutting, drying, tempering, bleaching, sterilization and extraction have been applied efficiently in the food industry. The aspects of food security and safety are important to study and analyze phenomena of food degradations, desired or not, on the texture and chemical composition of food processed by ultrasound.

Trends to develop are innovated not only on the techniques but also in original procedures, comprehension of the mechanisms of intensification of processing yield and activities and food degradation (desired or not) induced by ultrasound will be relevant as perspectives.

Prof. Pedro Cintas

Department of Organic and Inorganic Chemistry, University of Extremadura (UEX, Spain)

The safe and soft ultrasonic radiation, yet generating a vast amount of local energy owing to cavitational phenomena, has become an ideal companion in applied science, ranging from environmental issues to efficient preparation of micro/nanomaterials and gram-scale synthesis. What makes sonochemistry green is a combination of multiple effects that usually fulfill the 12 principles of green chemistry and/or chemical engineering. On improving selectivity and working under milder conditions, cleaner processes can be envisaged and hence less waste is created. Merging sonochemistry and green chemistry is a must, but if one misses a series of key points, the paradigm may also be flawed. Application of ultrasound to chemistry and (nano)technology requires to be green from scratch, paying attention to every aspect, among others solvent choice, purification steps and energy consumption, the latter often overlooked. It is expected that efforts like the present Handbook will be a stimulus in both education and research.

Prof. Dr. Juan Carlos Colmenares Quintero
Institute of Physical Chemistry of the Polish Academy of Sciences (Poland)

In recent years, a good number of methods are available for the preparation of an important group of nanomaterials called nanocatalysts. Nevertheless, the benefits derived from preparing nanocatalysts through unconventional approaches are very attractive from the green chemistry point of view. One of these promising new synthetic procedures that allow control over size, morphology, nanostructure and tuning of catalytic properties is sonication. Ultrasound-based procedures offer a facile, versatile synthetic tool for the preparation of nanocatalysts often inaccessible through conventional methods.

The unconventional ultrasound-based approaches of obtaining nanocatalysts are worthy of investigation and some of the reasons for this are: short times and environmentally friendly and cost-effective ways of nanocatalyst preparation, energy-economization and intriguing/promising nanocatalysts physico-chemical properties, among others. All these reasons fit perfectly in the scope of the green chemistry principles. Ultrasonic techniques open the doors for the preparation of the ideal heterogeneous catalysts with multiple integrated functional components that combine individual advantages to overcome the drawbacks of single-component catalysts.

The interesting use of ultrasonic irradiation in nanocatalysts synthesis is getting more and more valuable from both the fundamental and application points of view. Sonication is giving us this great opportunity as a real green and cost-effective methodology and is envisaged to hold great potential in the near future.

Prof. Giancarlo Cravotto

Dipartimento di Scienza e Tecnologia del Farmaco and NIS, Centre for Nanostructured Interfaces and Surfaces, University of Turin (Italy)

Both acoustic cavitation generated by ultrasound and hydrodynamic cavitation (HC) have proven to be important stepping stones towards process intensification in organic synthesis, placing sonochemistry among the elite of green chemical methods. Noteworthy environment-friendly application is the catalysis in aqueous biphasic systems under sonochemical conditions, in which cleaner products are easily obtained in higher yields with milder conditions and shorter reaction times. Sonochemical processes can reduce the formation of hazardous by-products, the generation of waste and also produce energy savings. Besides the low environmental impact, chemists can take advantage from recent progress in ultrasound and HC reactor engineering with new reliable flow systems. Complementary and synergistic effects have been found in the overlay between sonochemistry and other enabling technologies such as mechanochemistry, microwave chemistry and flow-chemistry. Recent studies provide a better understanding of correlated phenomena (mechanochemical effects, hot spots, etc.), and pave the way for emerging applications in green chemistry.

Prof. Micheline Draye

Laboratory of Molecular Chemistry and Environment (LCME), University Savoie Mont Blanc (France)

Sonochemistry indicates a physical process that uses sound waves of frequencies that are beyond human hearing, i.e., 16 kHz, to perform chemical reactions. Thus, above a sufficient power, sonochemical effects are observed in the liquid medium, which is submitted to the phenomenon of acoustic cavitation. Resulting bubbles concentrate the energy of the acoustic wave, and focus the timescales and length scales over which effects are observed: physical and/or chemical effects. In order to be green or sustainable, organic synthesis should meet some of the basic requirements suggested by the 12 principles of green chemistry, including atom and energy efficiency, less hazardous chemical synthesis, reduction of reagents requirements and use of catalysis to produce compounds that perform better or as good as the existing ones. Thus, if you ask me if ultrasound is able to serve the noble cause of green chemistry, I unambiguously answer: of course, it does! This method of physical activation allows chemists to carry out reactions without additional reagents or at least, with lower amount of reagents, for nanoparticles synthesis or phase transfer catalysis for example, addressing the atom efficiency and reduction reagents requirements. In addition, ultrasound affects the catalyst reactivity during catalysis by enhanced mass transfer and energy input, generation of unpassivated and highly reactive surfaces and production of particles of high specific surface, addressing the use of catalysis principle. An additional and unique benefit of the use of ultrasound is that it is able to switch a reaction pathway without changing the reagents that can address the principal of less hazardous chemical synthesis when these reagents are less hazardous themselves. Unfortunately, energy conversion is still a critical factor in the use of sonochemistry in the industry because of its low yield, which could affect the principle of energy efficiency. But, this problem is avoided when the use of ultrasound allows a sufficient decrease in the reaction time, addressing then the energy efficiency requirement!

Prof. (Em.) Aharon Gedanken

Department of Chemistry, Bar Ilan University, Ramat Gan (Israel)

Production of biofuels reduces the reliance on fossil fuels and in turn offers a green and sustainable pathway for the growing energy demands. Sonochemistry plays a vital role in the three major reactions, pretreatment, hydrolysis/saccharification and fermentation, involved in the production of second generation bioethanol using lignocellulosic feedstock as well as third generation bioethanol employing micro/macro algae as feedstock. Bioethanol is a promising alternative to fossil-based transportation fuels. Bioethanol production cost could be substantially reduced by the use of sonication based biomass pretreatment which is both fast and green and also enhances the subsequent hydrolysis process due to the reduction in structural rigidity. The improvement in the glucose (hydrolysis process efficiency) yield upon sonication is due to the disruption of chemical bonding. For instance, the rigid hydrophobic protein matrix and the amylose-lipid complexes surrounding the starch granules lead to higher glucose yield. Even lignin could be effectively removed by sonication from the complex lignocellulosic biomass, leading to the isolation of cellulose in high yields. Cellulose is vital for many technological applications in the current era, where there is a paradigm shift from fossil-based resources to renewables. Such a structural fractionation is the result of a combination of physical (acoustic cavitation) and mechanical effects occurring in the low frequency (16–100 kHz) ultrasonic treatment. Sonochemical irradiation is a green method for enhancing the enzymatic saccharification process which is a result of improved transport of enzymes, formation and subsequently collapse of cavitation bubbles causing increased accessibility of the reaction site of the biomass substrate to the catalytic species. Ultrasonic irradiation reduces the mass transfer constraints, structural rigidity of biomass, prevents the agglomeration of the catalytic species and there by accelerates the reaction under consideration leading to improved process efficiency. The process of fermentation of carbohydrates could also be substantially enhanced by appropriately modulating the frequency, power and the amplitude of ultrasonic waves and the irradiation time. Owing to these superior features, sonochemistry-based chemical processes could be adopted at the industrial scale leading to energy sustainability and green environment.

Sonochemistry was also employed in the transesterification of microalgae using a solid catalyst, SrO, however, although good yields of biodiesel were obtained, they were not as good as those obtained by microwave irradiation in terms of time of the reaction and the yield.

Dr. Parag Ratnakar Gogate
Institute of Chemical Technology, Mumbai (India)

Sonochemistry can give significant process intensification benefits driven by the chemical and physical effects induced by cavitation. In specific reference to green chemistry, the important benefits would be enhanced selectivity, use of less severe conditions, use of safer materials and intensified processing. Sonochemistry offers great promise for synthesis of high value low volume type materials including specialty chemicals, drugs with specific size distribution or nanocomposites with better characteristics. The other important areas of application are recovery of natural products, recovery of intracellular products as well as intensification of enzymatic reactions. Sonochemistry can also have a great opportunity in environmental protection especially based on the alternative mode of cavity generation, i.e., HC which can be more energy efficient. Identification of the controlling mechanisms for the specific application and deciding the optimum set of operating and geometric parameters would be essential for maximizing the benefits based on application of sonochemistry. Overall, sonochemistry is well established especially at laboratory scale with some demonstrations at pilot/commercial scale operations and combined efforts of chemists and engineers need to be directed at developing large scale designs possibly based on the principle of multiple frequency multiple transducers, which can help in eliminating the main problem of non-uniform cavitational activity distribution and erosion of transducers used for passage of ultrasound.

Prof. Hisashi Harada

Meisei University (Japan)

Sonochemistry has contributed in the field of green chemistry. Many kinds of pollutants are able to be eliminated from solution, such as ground water and sea water. Sonochemical reaction is effective even in cleavage of fluorine compounds, which is difficult to eliminate by other commonly used techniques. Synergy between sonochemical treatment and other degradation techniques is also effective. Not only the degradation of harmful compounds but also the sonopeeling and leaching of contaminants from solid materials are possible in solution at normal ambient. Sonopeeling may be useful for contaminated soil and sludge. To eliminate radioactive elements from contaminated soil, for example, sonopeeling is one of the important methods. Removing CO_2, namely greenhouse gas, is also an important matter for global environmental maintenance. Sonochemistry can contribute to help solve the problem. Furthermore, ultrasound-assisted organic synthesis should be able to make contribution to help escape from dependence on fossil fuels.

Dr. Georgios A. Heropoulos

Institute of Biology, Medicinal Chemistry & Biotechnology, National Hellenic Research Foundation (Greece)

Sonochemistry is one of the important tools that serve the principles of green chemistry and there have been many demonstrations of its successful application in this area. However, since the choice of energy source, e.g., conventional or non-conventional, plays a very important role and there is still room for further development of ultrasound applications in the areas of synthesis and processing both on a laboratory or industrial scale, this will result in a decrease in energy consumption and total cost by reducing reaction times, increasing product yields and avoiding solvents while reducing the need for waste treatment, finally giving better, cleaner products. Sonochemical techniques will continue to be used both in the destruction of hazardous products (e.g., chlorinated hydrocarbons, aromatic compounds, pesticides) or the cleaning of waste materials and in the synthesis of useful and safer products. For example, drug discovery and cosmetics are also very important areas where sonochemistry can contribute towards more efficient processes and cleaner products. Further studies concerning the application of different ultrasonic frequencies and the simultaneous use of two or more frequencies or the simultaneous combination of ultrasound with microwaves or light could also be beneficial. Other important areas where ultrasonic techniques are expected to play a major role are the decontamination of water, where the benefits from the application of ultrasound either alone or together with other types of irradiation have been demonstrated, and also the food industry where, in recent years, ultrasound has been found to be superior to conventional techniques in extraction, freezing, drying, etc., and in various aspects of processing and preservation. Both energy and food safety considerations should guarantee that these efforts will continue.

Prof. Jean-Yves Hihn

UTINAM Institute, UMR 6213 CNRS, University of Bourgogne Franche-Comté (France)

For many engineering situations, a natural inclination is to use materials that are intrinsically resistant, but the required properties are not necessarily all present in the same metal. Most of the substrates providing the better mechanical properties, for example high strength steels, also have poor resistance against corrosion. The solutions consist in plating processes used to coat an object with the desired metal by electrolytic or electroless modes. The final result can be considered for its "sustainable contribution" from one hand because it contributes to reduce the aging and obsolescence of a huge number of equipment, but on the other hand, the way to reach this objective uses chemicals often pointed out for their harmfulness and proven toxicity. Fortunately, it has been demonstrated that definite benefits may be gained by using ultrasound such as increased hardness and brightness, better adhesion to substrate, a finer grain and a reduced porosity and internal stress. Consequently, the intake of specific chemicals (ligands and organic surfactants) can be reduced or suppressed thanks to an ultrasonic irradiation. Ultrasound reduces the thickness of the diffusion layer as it produces cavitation close to the surface, resulting in microjetting, shock wave and convection. These three effects are known to enhance the mass transfer of electroactive species from the bulk solution to the electrode surface. Thus, under these conditions, more electroactive species are depleted at the electrode surface, which means an increase in cathodic efficiency. The latter is an important parameter in the electroplating industry; because it gets a Faradic yield deficit, either due to the electrolysis of the background electrolyte (e.g., production of hydrogen) or to the formation of a film at the electrode surface (e.g., metal oxide). Saving chemicals and energy with a little vibration input also confirms the "green chemistry label" of ultrasound in the surface treatment area.

Prof. Nilsun H. Ince

Institute of Environmental Sciences, Bogazici University (Turkey)

In my opinion, the following issues are very important in classifying the contributions of sonochemistry to the field of green chemistry:

1. Textile dyeing/finishing processes (sonication during the dyeing operation intensifies the extent of dye fixing on the fabric, thus reducing the dye concentration in the wasted dyebaths);
2. Water treatment (integration of ultrasonic systems to a conventional treatment plant increases the efficiency of coagulation/flocculation and disinfection processes);
3. Wastewater treatment (ultrasonic pretreatment of industrial process effluents before discharge to biotreatment operations reduces the toxic load by decomposing complex molecular structures);

4. Sludge disintegration (ultrasound is one of the most effective means of sludge disintegration providing "green" advantages such as low or no chemical consumption);

5. Surface cleaning properties of ultrasound contributes to green chemistry by prolonged life-time of materials, reduced rate of surface corrosion of metals and enhanced stability of catalysts (reducing their regeneration frequency (with chemicals) and waste as hazardous substances).

Dr. Robert Mettin

Drittes Physikalisches Institut, Georg August University Göttingen (Germany)

From my point of view, sonochemistry is still a "grey box" — not totally black, but several details of the physical phenomena that lead to the peculiar way of "stirring, compression and heating" by acoustic cavitation are not sufficiently well known yet to really understand what happens. It starts with the exact functional chain from electrical power and radiated sound waves *via* nucleation and activation of bubbles towards "hot spots of reaction" during their collapse — what distribution of bubble sizes and collapse strengths are generated under which circumstances? And it does not end with questions like: what happens exactly in the bubble collapse? Do we create shock waves in the gas and inhomogeneous temperatures? What mechanisms lead to mass transport between gas and liquid phase? How exactly does the erosion of the solid phase happen? And possibly most important for a successful implementation in green chemistry: How can sonochemistry be reliably controlled and optimized? Sensitivity and abundance of parameters can lead to a literally "complex system" that needs to be mastered. Significant progress has been made in the last decades, and more efforts and innovation can lead to a really deep understanding of sonochemistry in the future — and the well-directed use of cavitation for environment-friendly purposes. And for research, I am sure there is still a bunch of surprises ahead!

Prof. (Em.) Uwe Neis

Technical University Hamburg-Harburg (Germany)

In the past 20 years, ultrasonic technology has advanced considerably into the field of green chemistry and more precisely into the development of environment-friendly methods to produce renewable energy or to treat polluted liquid streams. In that context, ultrasound in its very broad frequency spectrum is often applied to replace commercial chemicals like methanol in municipal waste water treatment or to enhance the enzymatic activity of biological systems to turn organic wastes into biogas.

These two examples show a way ultrasound can help in the future finding more sustainable solutions where the threats to our planet are obvious: reduction/prevention of the pollution of air, waters and soils, production of renewable energy from organic wastes without greenhouse gas emissions.

Dr. Sergey I. Nikitenko
Marcoule Institute for Separative Chemistry (France)

In my opinion, sonochemistry is consistent with at least six of 12 principles of green chemistry developed by P. Anastas and J. Warner:

1. Waste Prevention
   Obviously, it is better to prevent waste than to clean it up. Sonochemistry allows one to minimize the concentration of reagents due to the significant acceleration of large variety of chemical processes and, consequently, to reduce the volume of secondary waste.
2. Less Hazardous Chemical Synthesis
   Ultrasonic treatment of the fluids generates the reagents (radicals and reactive molecules) *in situ*. Therefore, in some chemical processes driven with power ultrasound, the use of hazardous chemicals, such as strong oxidizers, can be avoided.
3. Safer Solvents and Auxiliaries
   Ultrasonic treatment has proven to be very efficient in solvent extraction processes of biologically active compounds. Due to that, in many systems, volatile harmful solvents can be replaced by water or aqueous solutions.
4. Design for Energy Efficiency
   Sonochemistry does not arise from a direct action of ultrasonic waves on molecules, but rather from acoustic cavitation: nucleation, growth and violent collapse of microbubbles in liquids submitted to ultrasound. The acoustic collapse generates transient extreme conditions

inside the bubble. In principal, each cavitation bubble can be considered as a microreactor providing high-energy processes at otherwise ambient conditions.

5. Catalysis

   Coupling of ultrasound with catalysts (sonocatalysis) has proven to be very efficient for organic synthesis and also for the degradation of organic pollutants in wastewater. The sonocatalytic effect is principally due to two phenomena: excellent dispersion of the catalysts in reaction medium and activation of the reaction sites at the surface of the catalyst.

6. Inherently Safer Chemistry for Accident Prevention

   The sonochemical processes are safe. The application of ultrasound does not lead to energy accumulation in treated solutions. Therefore, the sonochemical processes cease to happen when the ultrasound is off.

Prof. Sivakumar Manickam
University of Nottingham (Malaysia)

Sonochemistry is a true approach to green chemistry as it avoids toxic agents/solvents, high temperature, high pressure, longer period of reaction time which all lead to saving energy. In addition, the amount of solvents used in the process could be reduced. The immediate applications include extraction of active ingredients from natural products, homogenization and nanotechnology. Easy generation of nanoparticles and nanoformulations is feasible using this approach. Utilizing the radical species generated advanced oxidation is possible to a variety of difficult-to-treat pollutants. Looking at a diverse range of advantages it offers, it could be one of the potential approaches in green chemistry.

Prof. Tom Van Gerven
Process Engineering for Sustainable Systems (ProcESS),
Department of Chemical Engineering, KU Leuven (Belgium)

Besides the opportunities of new chemical pathways, I believe there is a lot of future in the use of ultrasound as an additional actuator in chemical processes in the sense that it is a non-contact energy form that allows to help in micromixing, degassing, erosion, hydroxyl formation, etc. New applications are shown in ultrasound-assisted crystallization (nucleation, crystal growth, fragmentation, polymorph selectivity). In the pharmaceutical industry, seeding is for example often practiced. However, external seeds can introduce defects or impurities in crystals. The use of ultrasonic cavitation can create heterogeneous nucleation sites without the need for seeds, thus consequently decreasing the risk of external contamination. Other opportunities are related to micromixing, which is difficult to achieve in stirred tank reactors, and even more in continuous flow reactors. Micromixing allows one to decrease the diffusion lengths and bring transport processes closer to their inherent rate, thus increasing mass transfer and resulting reaction rates.

References

1. T. J. Mason, *Ultras. Sonochem.* **2015**, *25*, 89–93.
2. R. Behrends, M. K. Cowman, F. Eggers, E. M. Eyring, U. Kaatze, J. Majewski, S. Petrucci, K.-H. Richmann, M. Riech, *J. Am. Chem. Soc.* **1997**, *119*, 2182–2186.
3. W. Lauterborn, W. Hentschel, *Ultrasonics* **1985**, *23*, 260–268.
4. Y. I. Frenkel, *Russ. J. Phys. Chem.* **1940**, *14*, 305–308.
5. M. A. Margulis, *Ultrasonics* **1985**, *23*, 157–169.
6. F. Lepoint-Mullie, T. Lepoint, R. Avni, *J. Phys. Chem.* **1996**, *100*, 12138–12141.
7. S. I. Nikitenko, R. Pflieger, *Ultrason. Sonochem.* **2016**, doi:10.1016/j.ultsonch.2016.02.003.
8. I. Hua, R. H. Hoechemer, M. R. Hoffmann, *J. Phys. Chem.* **1995**, *99*, 2335–2342.
9. M. E. Fitzgerald, V. Griffing, J. Sullivan, *J. Chem. Phys.* **1956**, *25*, 926–933.
10. K. S. Suslick, D. A. Hammerton, R. E. Cline, *J. Am. Chem. Soc.* **1986**, *108*, 5641–5642.
11. J. P. Bazureau, M. Draye, *Ultrasound and Microwaves: Recent Advances in Organic Chemistry*, Research Signpost, Kerala, India, 2011.
12. T. Y. Wu, N. Guo, C. Y. Teh, J. X. W. Hay, Challenge and recent developments of sonochemical processes, In: *Advances US Technology for Environmental Remediation*, SpringerBriefs in Molecular Science, Netherlands, 2012, 109–120.
13. P. T. Anastas, J. C. Warner, *Green Chemistry: Theory and Practice*, Oxford University Press, Oxford, 1998, 11–54.

14. P. T. Anastas, N. Eghbali, *Chem. Soc. Rev.* **2010**, *39*, 301–312.
15. M. Sillanää, T.-D. Pham, R. A. Shrestha, *US Technology in Green Chemistry*, SpringerBriefs in Green Chemistry for Sustainability, Netherlands, 2011.
16. G. Cravotto, P. Cintas, *Chem. Sci.* **2012**, *3*, 95–307.
17. S. L. Fegade, J. P. Trembly, *Ultrason. Sonochem.* **2015**, doi:10.1016/j. ultsonch.2015.04.007.
18. P. Cintas, *Ultrason. Sonochem.* **2016**, *28*, 257–258.
19. R. A. Al-Juboori, T. Yusaf, L. Bowtell, V. Aravinthan, *Ultrason. Sonochem.* **2015**, *57*, 18–30.
20. D. F. Rivas, P. Cintas, H. J. G. E. Gardeniers, *Chem. Commun.* **2012**, *48*, 10935–10947.

Index